Routledge Revivals

Assessing Surprises and Nonlinearities in Greenhouse Warming

In 1992, Resources for the Future conducted a workshop concerning the issues of global climate change. This title, originally published in 1993, is a collection of the revised versions of the papers commissioned for the workshop with an added introduction and overview. Each paper emphasises the potential nonlinearities or surprises in physical effects caused by humans loading the atmosphere with greenhouse gases and examines how shifts in the natural environment from climate change may affect human well-being. This collection is a valuable resource for any student interested in environmental studies and climate change issues.

Assessing Surprises and Nonlinearities in Greenhouse Warming

Proceedings of an Interdisciplinary Workshop

Edited by
Joel Darmstadter and Michael A. Toman

RFF PRESS
RESOURCES FOR THE FUTURE

First published in 1993
by Resources for the Future, Inc.

This edition first published in 2016 by Routledge
2 Park Square, Milton Park, Abingdon, Oxon, OX14 4RN
and by Routledge
711 Third Avenue, New York, NY 10017

Routledge is an imprint of the Taylor & Francis Group, an informa business

Publisher's Note
The publisher has gone to great lengths to ensure the quality of this reprint but points out that some imperfections in the original copies may be apparent.

Disclaimer
The publisher has made every effort to trace copyright holders and welcomes correspondence from those they have been unable to contact.

A Library of Congress record exists under LC control number: 93005534

ISBN 13: 978-1-138-95371-0 (hbk)
ISBN 13: 978-1-315-66719-5 (ebk)

ASSESSING SURPRISES AND NONLINEARITIES IN GREENHOUSE WARMING

PROCEEDINGS OF AN INTERDISCIPLINARY WORKSHOP

JOEL DARMSTADTER
AND
MICHAEL A. TOMAN
EDITORS

Resources for the Future
1616 P Street, NW
Washington, DC 20036
May 1993

Printed in the United States of America

Published by Resources for the Future
1616 P Street, NW; Washington, DC 20036-1400

Library of Congress Cataloging-in-Publication Data

Assessing surprises and nonlinearities in greenhouse warming:
    proceedings of an interdisciplinary workshop/Joel Darmstadter,
    Michael A. Toman, editors.
            p.   cm.
        Includes bibliographical references.
        ISBN 0-915707-71-3: $25.00
            1. Climate changes—Congresses. 2. Global warming—Congresses.
        3. Greenhouse effect, Atmospheric—Congresses. 4. Nonlinear theories—Congresses.
    I. Darmstadter, Joel, 1928-   . II. Toman, Michael A.
    QC981.8.C5A89   1993
    551.6—dc20
                                                                        93-5534
                                                                        CIP

Contents

Preface and Acknowledgments

Society has a great interest in the risks posed by global climate change, both in taking steps to avoid unacceptable damage and in avoiding unwanted or ineffective mitigation measures. Unfortunately, this interest is not matched by available knowledge. Physical science aspects of climate change—how much warmer, wetter, or drier, and how variable climate might be in different regions with different atmospheric concentrations of greenhouse gases—are uncertain and are likely to remain so for many years to come. And even if one posits particular climatic shifts, the ecological, social, economic, and other human consequences are elusive.

These uncertainties pose sharp dilemmas for determining research priorities and for individual and social decision making. On what basis should society make decisions about the amounts and types of mitigation or adaptation efforts that should be undertaken? What values are put at risk by climate change? What kinds of uncertainties most cloud the picture? In particular, how significant are potential nonlinearities or "surprises" in physical or human effects? What research directions and other steps are appropriate for shrinking these uncertainties?

In March 1992, Resources for the Future conducted an interdisciplinary workshop centering on these issues. The workshop was built around commissioned papers written by leading experts in several relevant natural and social science fields. Revised versions of these papers, along with an added paper (by Clark and Reid) and an introduction and overview, constitute the present publication.

As with any volume based on conference papers by many authors, the contributed papers herein differ in tone, style, method, scope, and depth of coverage. There are gaps in some presentations and disagreements among authors in some conclusions. Rather than attempting as editors to reconcile all these issues, we have sought to produce a more expedited volume in which the authors largely are left to make their own arguments. Given the strong current interest in climate-related nonlinearities and uncertainties, our hope is that the volume will make a more timely contribution even with some elements of exposition and interpretation left unresolved.

We wish to acknowledge the financial support of the Electric Power Research Institute (EPRI) in sponsoring the workshop and facilitating the preparation of this volume. We are grateful to Hung-po Chao and Stephen Peck of EPRI for their guidance—beginning with planning for the workshop and extending to the publication of the present report. We and the authors benefited from the large amount of lively and thoughtful discussion at the workshop, both within and across disciplinary lines. We especially want to thank William Cline, Hadi Dowlatabadi, Jeff Hyman, and an anonymous reviewer for their helpful comments and suggestions as this volume was being assembled. Of course, as always, responsibility for the contents of each chapter rests entirely with its authors. We owe a large debt to Kay Murphy for her efforts in facilitating organization of the workshop and preparing the manuscript. In addition, we wish to thank Dorothy Sawicki, of RFF's publications staff, for overseeing the editorial processing of the manuscript; and Diane Kelly for her work in designing and producing the book.

1

Nonlinearities and Surprises in Climate Change: An Introduction and Overview

Joel Darmstadter and Michael A. Toman

It is only a slight overgeneralization to say that current debates about the magnitudes and consequences of human-induced climate change are debates about what is not known. The emphasis is placed on what <u>could</u> (with varying degrees of confidence) occur as humankind continues to load the atmosphere with greenhouse gases. Particular concern is expressed about effects that are in some sense unexpected, in that they are not natural extensions of existing trends or within the limits of predictive models. An important subtext in this concern about surprises is the question of nonlinearities—responses of natural or socioeconomic systems that are disproportionate to the changes in stimuli and may threaten the adaptive capacities of the systems in question. Nonlinear and unexpected impacts complicate an already difficult problem of valuing effects that span generations and involve large spatial scales with substantial nonmarket assets as well as market goods.

In March 1992, Resources for the Future held a conference to examine these issues from both natural science and economic perspectives. The goals of the workshop were (1) to examine the existing state of knowledge regarding surprises and nonlinearities in natural and socioeconomic systems confronted with climatic change; (2) to promote interdisciplinary exchange of information regarding these issues; (3) to distill insights that could be communicated to a wider audience; and (4) to draw conclusions about productive directions for further research on climate change.

The papers review a wide range of topics, including the following:

- How continued greenhouse gas emissions may affect climatic conditions.
- How climate change may affect agriculture, the archetypical managed ecosystem, and less managed ecosystems.
- How various shifts in the natural environment from climate change may affect human well-being, with particular concern about damages that may vary nonlinearly with climate change.
- How improved information about climate change and its consequences may have value in guiding decision making.

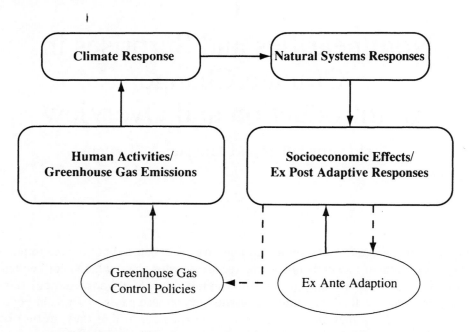

Figure 1. Climate change and its interaction with natural, economic, and social processes. See text discussion.

All of these questions arise in connection with a complex set of interdependent natural and socioeconomic processes. Figure 1 provides a simple representation of these processes. A variety of human activities give rise to greenhouse gas emissions such as carbon dioxide from fossil fuel production and methane from agriculture and energy systems. These emissions could lead to climatic responses such as changes in average temperature and rainfall and changes in variability of climatic conditions. Altered climatic conditions could affect managed and unmanaged ecosystems (changes in agricultural yields, streamflows, habitat ranges), which in turn could have a number of socioeconomic consequences (declines in real income, population dislocation, damages to infrastructure and environmental capital).

After the fact, individual human actors would respond to the socioeconomic consequences of climate change in ways that can be expected to alter subsequent emissions. This response could be negative or positive; e.g., the response to greater dryness could be increased energy use to irrigate agriculture but less use of air conditioning. Of course, people also can and presumably will engage in before-the-fact adaptation based on judgments about the consequences of business as usual; more drought-resistant species can be cultivated, and construction can be pushed back from the shoreline. In addition to adaptation, societies can undertake mitigation policies that seek to reduce future climate change by curbing greenhouse gas emissions.

How do issues of nonlinearity and surprise arise within such a schematic representation? Fundamentally, the concept of nonlinearity is a mathematical abstraction: we describe the relationship between two defined entities x and y to be nonlinear if there is a well-defined relationship between them and that relationship has a nonconstant slope. This can occur because of smooth changes in slope or because of kinks or discontinuities (both of which can reflect thresholds in observed responses). All of the subsequent papers in this volume illustrate such effects. Of course, for the abstraction to be useful, care must be taken in specifying the relationship: an exponential relationship between x and y is a linear relationship between their logarithms.

If we had complete knowledge of the physical and social systems represented in figure 1, nonlinearities would be important features to consider in gauging adaptation and mitigation strategies, but nothing more. However, we do not and (almost surely) never will have such complete information. Uncertainty is an inherent part of the process by which human beings confront their environment,[1] while knowledge allows us to revise our expectations over different states of the world (Morgan, Henrion, and Small, 1990).

Seen from this point of view, surprise about one or another aspect of climate change is an after-the-fact reaction to an observation or new scientific finding that, in some sense, lies well outside our range of expectations—for example, a highly unexpected decline in agricultural productivity or biodiversity from shifts in rainfall or temperature. If the phenomenon is (or is expected to be) persistent rather than transient or anomalous, then the surprise also is an indication that our prior expectations about the natural and human systems in question need revision. Here we see how nonlinearity and surprise interact: given our tendency both scientifically and intuitively to extrapolate observed trends, a nonlinearity not incorporated into our prior information and set of beliefs is likely to engender surprise.

The problem is that often we don't like surprises because they threaten to be costly. Mitigation and adaptation strategies conditioned on one set of expectations must be revised, and this may be quite costly. Moreover, when unexpected consequences are adverse and effectively irreversible (permanent ecosystem disruption or coastal flooding), we must simply live with the consequences.

One appropriate response in these circumstances is research that looks at the broad outlines of interdependent systems as well as individual key parameters, in order not only to reduce uncertainty generally but also to reduce uncertainty about potential nonlinearities. One purpose of the papers in this volume is to highlight where these uncertainties may lurk and how to better understand them. We return at the end of this Introduction to some brief observations about future research directions. Another response to the combination of nonlinearity, uncertainty, and irreversibility is a more aggressive approach to mitigation so that future options are not foreclosed. While the prudence of this strategy holds in principle, the empirical issue of how aggressively we should mitigate remains hotly disputed.

CONTENTS OF THE VOLUME

In the first of three papers devoted to natural science dimensions of the greenhouse problem, *Norman Rosenberg* contributes a background survey of the state of knowledge concerning climate change. Drawing on a wide-ranging body of research, Rosenberg, like numerous other scientists, finds that rising atmospheric concentrations of greenhouse gases may well lead to significant warming of the atmosphere during the next century. This is a prospect that prompts hime—invoking prudence—to stress the desirability of significantly slowing the increase of greenhouse gas emissions. The principal gases of concern (and their estimated contribution to total "radiative forcing" during 1980-90) are carbon dioxide (CO_2), 55%; methane (CH_4), 15%; nitrous oxide (N_2O), 6%; and the chlorofluorocarbons (CFCs), 17%. Rosenberg cites specifically the judgment of scientists associated with the Intergovernmental Panel on Climate Change (IPCC) that "an equilibrium climate change due to a doubling of CO_2 or its radiative equivalent due to all the greenhouse gases will warm the lower atmosphere (troposphere) and cool the stratosphere and that global average tropospheric warming will range between 1.5 degrees Celsius (°C) and 4.5°C with 2.5°C as the 'best guess'." Included in Rosenberg's discussion is the question of the extent to which the prospects for detrimental consequences of greenhouse warming may be at least partially attenuated. Prominent examples in agriculture are heightened photosynthetic stimulus on crop growth and enhanced water use (topics further detailed in Paul Waggoner's paper in this volume).

Notwithstanding the emerging scientific consensus regarding the prospects for a several-degree increase in mean global temperature in the course of the next century, the recurrent note of caution in Rosenberg's treatment highlights the many aspects of climatic change about which our knowledge remains deficient. Indeed, Rosenberg's discussion provides cogent examples that help to differentiate several of the taxonomic concepts laid out earlier in this Introduction—particularly questions of uncertainty, nonlinearity, and surprise. He points out how general circulation models (GCMs)—the primary means of relating global temperature change to atmospheric greenhouse gas concentrations—remain, in their present state of development, limited predictive tools. Thus, much uncertainty surrounds the "best guess" estimate of 2.5°C warming alluded to earlier. The low (1.5°C) and high (4.5°C) increments cited above reflect the potential nonlinear consequences of positive or negative feedbacks triggered by an initial rise in temperature. For example, warming in high latitudes may cause the release of methane trapped in hydrates—a positive feedback accentuating global warming. (The estimated feedbacks are themselves subject to uncertainty ranges.) The drier soils resulting from a warmer earth are a sink for methane, thereby lowering its atmospheric concentration and causing a negative feedback. One of the most complex and potentially significant secondary consequences of greenhouse warming involves what happens to cloud cover; here the feedback effect may be positive or negative.

Rosenberg's discussion also illustrates the phenomenon of surprise. Until recently, the CFCs were viewed as a powerful constituent of the greenhouse effect, their diminishing future importance only due to the phaseout dictated by the Montreal Protocol and its subsequently strengthened provisions. But, in what appears to have come as a surprise to the scientific community, it "now appears . . . that the ozone losses attributable to CFCs could actually lead to atmospheric cooling rather than warming." As Rosenberg observes, "CFCs illustrate well the rapidity with which new scientific findings relating to greenhouse warming replace not-so-old ones."

Rosenberg's overview of greenhouse warming is followed by *Paul Waggoner*'s paper addressing nonlinearities and surprises in the impact of climate and weather on agriculture. The scope of the paper extends to farming in general as well as selected key agricultural topics, such as drought and water runoff.

Many examples presented in Waggoner's discussion make it clear that the linkage between weather and farming abounds with nonlinearities. For instance, for many crops the pace of germination of seeds proceeds in nonlinear fashion beyond a threshold temperature of 4°C. Another example involves nonlinearity when leaf enlargement is plotted against moisture. And there are nonlinear linkages between CO_2 concentrations and photosynthesis in the growth of certain classes of crops. Finally, there are nonlinear characteristics in the maturation stage of an organism's weather-related flowering and grain-filling stages.

Clearly, weather produces both complex linear and nonlinear effects throughout a plant's life cycle from embryo to maturity; and, aside from responses endogenous to the organism, climatic effects on pests and frost further complicate the story. Unraveling "net" agricultural impacts of climatic change thus is no simple matter. Waggoner does, however, provide some insights into the extent to which agriculture as a whole, or a major farm sector such as wheat production, responds linearly or nonlinearly to climatic influences. He suggests, for example, that in contrast to the nonlinearity characterizing the drought years of the 1930s for a wheat-producing Kansas county, adaptation can mute nonlinear responses over a longer period of time. In that context, he cites evidence relating to the 60-year northward migration of hard red winter wheat in adaptation to a changing regional climatic regime. But in turning from empirical evidence about effects of climatic change to a broader consideration of agricultural response under different scenarios, Waggoner acknowledges that there are also nonlinearities in the incremental effects of adaptation, whose effectiveness could be limited by the introduction of species with lower yields in the interest of greater resilience to long-term climate change.

Are assumed changes in climate likely to produce nonlinear agricultural surprises? Waggoner believes that surprise can be lessened and adaptation prepared for if more attention is directed to studying nonlinearities at the vulnerable geographic margins of climatic zones. But when all is said and done, simply showing nonlinearities does not resolve a planner's dilemma. What is called for, in Waggoner's view, and as he diagrammatically illustrates, is a practical threefold strategy: neglecting inconsequential climate change, neglecting, as well extremely unlikely events that it would be extremely costly to deal with, and thus concentrating on a "middle ground" requiring agricultural adaptation.

In contrast to Paul Waggoner's concern with the consequences of climatic change for one vital sector of human-directed economic activity, *James Clark* and *Chantal Reid* review a number of implications of greenhouse warming for a variety of—largely unmanaged—ecosystems.[2] The analysis emphasizes the interlinked concepts of sensitivity, feedbacks, and nonlinear responses, and proceeds in two major stages. First, the authors take up the principal processes and characteristics that play a key role in ecosystem sensitivity to climatic change. The treatment considers different ecological regimes rather than any given area of the world. Hydrology, soil fertility, nutrient cycling—and the role of these factors in the carbon cycle—are among the topics discussed. Second, the authors examine three particular ecosystems in North America (arctic regions, temperate forests, and salt marshes) in terms of the factors introduced in the first part of their review.

Not surprisingly, given the interplay of the many forces that constitute an ecosystem, the picture that emerges is one of great complexity—scarcely helped by the fact that GCMs of climatic change are poorly equipped for predictions at the geographic scale associated with one or another ecosystem. For example, predicting climate-induced precipitation changes for an ecosystem presupposes GCM modeling that is spatially much more discriminating than the minimum grid cells currently employed.

Notwithstanding the complexity of their subject, Clark and Reid provide valuable insights and informed judgments on a range of prospective ecological impacts of greenhouse warming. To cite just one example, increased length of the growing season and moisture availability are likely to result in higher productivity of organic matter in regions where these elements are currently limiting. On the other hand, CO_2 fertilization effects—examined also in the papers by Rosenberg and Waggoner—are more problematic. For one thing, many plants are not carbon limited; also, the stimulative effect of new tree growth may increase mortality under closed-stand conditions, thus attenuating the fertilization benefit.

In these and other instances, the authors delineate both positive and negative impacts—applying those terms in a directional rather than normative sense—and identify nonlinear phenomena where these appear to be in prospect. High latitudes, particularly arctic ecosystems, appear to be especially prone to nonlinear responses to climatic change, with temperature playing a more decisive role than direct CO_2 effects. Temperate forest responses defy so straightforward a characterization. Such things as soil moisture-holding capacity, shallowness of soils, seasonal temperature variability, competition among species, carbon losses from organic matter—to name just some factors at work—complicate the task of analysis. Although the authors willingly speculate on the outlook for nonlinear impacts, they emphatically caution against attempts to quantify change in environments for which no analog exists and where human disturbance may be significant.

Uncertainty about the potential damages to human and natural systems, discussed in physical dimensions in the first three papers, forms the point of departure for the ensuing three papers cast within the perspective of economic analysis. In the first of these papers, *Stephen Peck* and *Thomas Teisberg* focus on the implications of uncertainty about the damage function—specifically, the possibility of its nonlinear character. These authors undertake three major tasks: (1) sensitivity analyses designed to draw out implications for optimal policy of alternative damage function assumptions; (2) calculations designed to show how information about the level and curvature of damage may produce benefits in the form of accelerated resolution of uncertainties; and (3) an assessment of the implications of damage function nonlinearity for the value of information. Their focus in task

3 is on a scenario that drives much climatic change analysis and public policy discussion: doubling of the CO_2-equivalent atmospheric concentrations and the associated global mean temperature increase. In addressing these three questions, Peck and Teisberg exploit the potentialities of a model termed CETA (carbon emissions trajectory assessment), embracing worldwide economic growth, energy consumption, energy technology choice, global warming, and global warming costs.

With respect to the first question, Peck and Teisberg find that optimal policy is substantially more sensitive to the <u>degree of nonlinearity</u> in damages—i.e., to the rate of change or power characteristic of the damage function—than to the <u>level</u> of damages at a specified temperature increase. The authors' second topic gives rise to a related finding: that the value of information about damage function nonlinearity exceeds the value of information about the damage function level. Finally, exploration of the third question supports one's intuitive judgment that the value of information concerning the rate of increase in warming is enhanced if the damage function exhibits a "highly nonlinear response to temperature change." This underscores the importance of resolving uncertainties about the extent to which temperature change is a function of greenhouse gas concentrations.

Peck and Teisberg conclude from this analysis that the degree of nonlinearity is a key uncertainty affecting optimal policy choice and that their results highlight the central importance of research designed to estimate the cost of climatic damage, with particular emphasis on the nonlinear path exhibited by that damage.

Both of the remaining papers of this volume revolve around the desirability of moving beyond one of the few efforts—that by Nordhaus (1991)—to estimate what magnitude of greenhouse gas abatement may be economically justified by the prospective dollar measure of damages avoided.[3] *Gary Yohe*'s paper illustrates in the specific case of sea-level rise how one might analyze, in economic terms, the prospective vulnerability stemming from climatic change. This is clearly a task one undertakes with some trepidation, recognizing that the uncertainty with which the future effects of global change phenomena are currently viewed introduces analytical complexity even before consideration of nonlinear and dramatic surprises is added to the calculus. Yohe nevertheless believes that evidence of a substantial degree of shared concern with low-probability, significant-impact outcomes warrants investment in research that could strengthen both understanding of, and possible ways of dealing with, the climate change dilemma. It is, in short, not very helpful to view extreme events as unthinkable.

Yohe's approach, built on the sea-level case study, involves employing . . ."the Nordhaus analytical structure to explore the degree to which elevated estimates of damages which are consistent with current and coincident subjective views of future circumstances for the United States might enlarge the theoretically justified efficient response." The resultant estimates of damages and their nonlinear consequences increase the baseline marginal damage of Nordhaus by more than 160% and, correspondingly, raise the economically efficient reduction in U.S. CO_2 emissions from 6% below the baseline to approximately 15% below. Yohe readily acknowledges a substantial degree of simplification in such quantification; yet, as he notes, the "lesson that systematic inclusion of possible surprise events and nonlinear damages should make us more cautious in the protection of our health and well-being should not . . . be dismissed as the consequence of oversimplification."

Ideally, economic assessment of damage due to climatic change should apply to both resources denominated in market value (such as in agriculture) and the many ecological resources that largely lack any sort of basis for monetary estimation. *Anthony Fisher* and *Michael Hanemann* provide a critical review of each of these challenges, noting that while the first task is unquestionably the more tractable one, it is in itself a far from straightforward undertaking. They illustrate this point by examining two prominent efforts to assess the economic dimensions of climatic change in the United States—analyses by Nordhaus (already referred to) and by the U.S. Environmental Protection Agency (EPA).

The discussion points up numerous problems that deserve airing. For example, quite apart from EPA's understandable hesitancy in tackling impacts on nonmarket assets and human health, Fisher and Hanemann fault the agency for neglecting economic impacts associated with changed patterns of water availability for agricultural, municipal, and industrial uses and for ignoring impacts on economic infrastructure generally. This incompleteness in coverage appears to reflect the inadequacy of existing economic models to address some important sectors vulnerable to the effect of greenhouse warming.

The authors raise questions about several other aspects of EPA's and Nordhaus's valuation efforts: Is proper account taken of diminishing land availability as affected populations (say, in coastal areas) are forced to move inland? Pressures on land availability could clearly signify nonlinear response patterns. Is the valuation of affected infrastructure biased too greatly toward historic rather than replacement cost? Are there subtle linkages between private property and nearby public infrastructure that ought to be reflected in estimates of impacted property values? What about premature abandonment of capital rendered obsolete by climatic impact? As the public debate over the greenhouse issue and its economic dimensions widens, such topics are likely to receive increased attention.

In turning to the dilemma of quantifying damages to natural ecosystems, Fisher and Hanemann lean to the view of ecologists who assert that the costs are likely to be substantial, that in certain respects—e.g., a "worrisome loss" of biodiversity—the trend may be an accelerating one, and that the results are likely to be irreversible. The authors also press for analysis which, like Cline's (1992), contemplates a time horizon substantially more distant than that conventionally equated with a doubling of atmospheric CO_2 concentrations. They argue that such a longer perspective may reveal departures from equilibrium states that are "explosive," or, in the taxonomy of this volume, nonlinear.

This observation leaves open the question of probabilities versus possibilities, and in terms of the focus of Fisher and Hanemann's paper, raises the challenge of incorporating such contingencies into economic valuation. In their concluding discussion, the authors assume that the likelihood of nonlinearities and sharp discontinuities in the warming damage function (not to mention the greater or lesser "option value" associated with the preservation of natural capital generally in the face of irreversible damages) is sufficiently high to justify a slowing of emissions. This could allow time for the scientific inquiry that would make assumptions about these effects more robust. Thus, while Fisher and Hanemann's conceptual framework does not contradict that of Peck and Teisberg, Fisher and Hanemann make a stronger normative argument for greenhouse gas abatement policies.

IMPLICATIONS FOR FUTURE RESEARCH

The papers in this volume are diverse and are as much concerned with cataloging knowledge gaps as with drawing specific conclusions. Thus, it was too much to hope that a specific agenda for research, much less a set of specific guidelines for decision making, would emerge from the conference. Nevertheless, several broad observations are suggested by the papers and discussion at the workshop.

It is clear, first, that research on damages from climate change deserves a high priority. Lots of work has been done on the potential costs of slowing climate change by abating greenhouse gas emissions. While considerable disagreement attends the estimates, it is possible to bound the order of magnitude of abatement cost and to identify reasons for disagreement among estimates (e.g., long-term versus short-term analysis, different assumptions about technology and fuel substitution possibilities). Even this limited level of confidence eludes us in considering potential damages of climate change. The natural and socioeconomic components remain highly uncertain (in the sense defined above), as do their interactions and the possibilities for constructive adaptation.

Contrary to what some skeptics contend, we believe that it is premature to judge future climate change to be a minor environmental and social issue. Collectively, the papers in this volume suggest that damages could be very substantial. In terms of research priorities, they indicate the importance of better understanding socioeconomic consequences, not just physical effects. However, the analyses presented here also highlight that nonlinearity in damages may be at least as important as expected damages for some postulated extent of climate change. Point estimates in this context have limited use; it is necessary to consider ranges of estimates from a variety of model structures.

To understand nonlinearity (or discontinuity) in damages requires, as the natural scientists in the volume point out, analysis of natural thresholds that could trigger significant socioeconomic effects. Just as important, as emphasized by economists' contributions, are an understanding of nonlinear socioeconomic responses and the identification of significant obstacles to human adjustment whose alleviation could make damages less nonlinear. Such information acquisition may not be the only prudent insurance in the face of the uncertainties posed by climate change, but it clearly has high social value.

NOTES

1. We take no view here on the abstract issue of whether natural systems are inherently stochastic or whether that is a property of incomplete information; for our purposes, the implications are the same.

2. This paper was commissioned subsequent to the workshop.

3. The major study by Cline (1992), which sets its sights substantially beyond the scope of Nordhaus's analysis, appeared at just about the time of the workshop on which this volume is based. Cline's work, therefore, does not receive the attention it deserves in this volume, although it received significant discussion during the workshop.

REFERENCES

Cline, William R. 1992. *The Economics of Global Warming* (Washington, DC: Institute for International Economics).

Morgan, M. Granger, Max Henrion, and Mitchell Small. 1990. *Uncertainty: A Guide to Dealing with Uncertainty in Quantitative Risk and Policy Analysis* (New York, NY: Cambridge University Press).

Nordhaus, William D. 1991. "To Slow or Not to Slow: The Economics of the Greenhouse Effect," *The Economic Journal*, July, pp. 920-937.

2
Facts and Uncertainties of Climate Change

Norman J. Rosenberg

We must first recognize that climate is always changing. There is evidence of this in tree rings (Fritts, 1966), in the pollen conserved in lake sediments (Bernabo and Webb, 1977), in the concentrations of gases trapped in Greenland and Antarctic icecaps (Stauffer et al., 1985; Neftel et al., 1985). Glacial moraines and other landforms that could only have been created under climatic conditions very different from the present ones also testify to climate change. The concern of this conference is not with those natural climate changes that occur over millennia, but rather with climate changes induced by human activities that alter land use and the composition of the atmosphere. Such changes could occur within the coming decade and century or could possibly be occurring already. We fear that climate changes of the latter kind could be both severe enough and rapid enough to cause serious dislocations and require difficult adaptations.

This paper deals with a limited number of topics: how climate changes can occur through changes in land use and atmospheric composition; how increases in the concentration of radiatively active trace gases can lead to climate warming; how warming may alter other aspects of climate; and how the increase in atmospheric carbon dioxide (CO_2), one of the major greenhouse gases, can affect plant growth. An attempt is also made to illustrate how certain of the physical and biological phenomena discussed are linked and to illustrate the uncertainties in our current understanding of these linkages.

This paper relies very heavily on the comprehensive report of the IPCC Group I (Intergovernmental Panel on Climate Change [IPCC], 1990) and on the results of research published since that report was issued.

I begin with a "mini-primer" on climate change.

A MINI-PRIMER ON CLIMATE AND CLIMATE CHANGE

Climate is the mean or average condition of the atmosphere—in other words, the mean or average weather. More technically, climate is the statistical description of the mean state, including the variability of the atmosphere, ocean, ice, and land surface in a specified period of time (National Academy of Sciences [NAS], 1975). Weather (the motions of the atmosphere) is generated by the spin of the earth on its axis and by the differential distribution of solar radiation on the earth's surfaces. The differences in distribution are due to latitude, season, turbidity of the atmosphere, and cloud cover.

The atmosphere is an envelope of gases that covers the planet to a thickness of about 35 kilometers (km). At the top of the atmosphere incoming solar radiation is fairly constant. There, a surface normal to the rays of the sun would receive solar radiation at a flux density of about 1,370 watts per square meter (W m^{-2}). (The mean flux density of solar radiation at the top of the atmosphere for the earth as a whole is 340 W m^{-2}.) On entry to the atmosphere, a portion of this radiation is reflected (see figure 1) and a portion is absorbed, but the largest amount reaches the surface of the earth because the gases of the atmosphere are quite transparent to the visible radiation (0.4 to 0.7 micrometers) which predominates in the solar spectrum. The atmosphere, far cooler than the sun, emits radiation in the longwave, or thermal, band (approximately 8.0 to 14 micrometers) and some of this radiation is also directed toward earth.

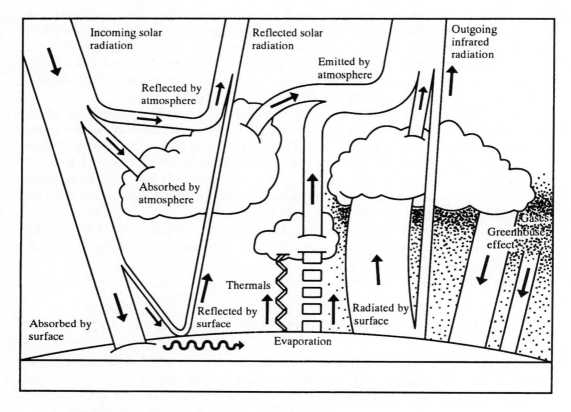

Figure 1. Schematic illustration of the earth's radiation and energy balances (Source: Figure 2–1 in S.H. Schneider and N.J. Rosenberg, 1989).

A considerable fraction of the shortwave solar radiation that reaches the surface is reflected in the direction of space. The remainder of the shortwave radiation and most of the impinging longwave radiation is il absorbed by the land, vegetation, and bodies of water. This radiant energy warms the surface, which in turn warms the air that comes in contact with it. The absorbed radiation also provides the energy that evaporates water. Vapor carries this energy in the form of latent heat into the atmosphere. The energy not dissipated as already described is transferred back to space as longwave thermal radiation. However, natural constituents of the atmosphere—water vapor, CO_2, methane (CH_4), nitrous oxide (N_2O), and ozone (O_3)—and some manufactured substances, such as the chlorofluorocarbons (CFCs), are partially opaque to the longer wavelength thermal radiation and trap a portion of it. A portion of the energy absorbed by these so-called greenhouse gases is retained in the lower layers of the atmosphere, raising its temperature. Were it not for this natural "greenhouse effect" the atmosphere would be about 33°C cooler on average than it is.

Thus the radiation balance of the earth-atmosphere system is determined by the exchange of shortwave visible and longwave thermal radiation. This balance can be altered (and climate will be changed as a result) in many ways. Incoming solar radiation varies over time with changes in solar luminosity. The sun has grown in size and brightened by about 25% over the last 4 billion years (Schneider and Londer, 1984). The angle of incidence of solar radiation varies with changes in the orbital relations between sun and earth and with changes of the earth's tilt on its axis. Changes in orbit and tilt occur with regularity but the cycles for such changes are long—varying between roughly 20,000 and 100,000 years. On much shorter time scales the transparency of the atmosphere to solar radiation varies with volcanic activity,[1] with aridity and consequent dustiness, and with changes in cloud regime.

A portion of the incoming solar radiation is reflected back to space by the atmosphere and clouds. A portion of that reaching the surface is also reflected. The albedo (reflectivity) of earth's surfaces varies: for forest, it is about 0.18 (18% reflectivity); for grassland, about 0.24; for a lake, about 0.10; for desert, about 0.50 (Rosenberg et al., 1983). Thus changes in land use, if on a large enough scale, can alter global climate by altering the amount of solar radiation retained at the surface. Land use change can also alter temperatures at the earth's surface and, hence, the emission of longwave thermal radiation, which varies with the fourth power of the surface temperature.

The second mechanism by which humankind is able to alter climate is through changes it causes in the composition of the atmosphere. The role of "greenhouse," or radiatively active, trace gases has already been mentioned. Increases now occurring in the concentrations of the natural greenhouse gases and industrial compounds such as CFCs that have greenhouse properties have the potential for further warming of the atmosphere. Atmospheric radiative transfer models indicate that a doubling of the greenhouse gas CO_2 would increase the global average radiative flux at the top of the atmosphere by 4.4 Wm^{-2}—about 1.3% of the mean flux at the top of the atmosphere. Since other greenhouse gases—CH_4, N_2O, and the CFCs—also contribute to this potential for altering radiative flux, the current understanding of their individual roles is reviewed in the following section. Other substances that derive from industrial and/or agricultural and biological systems can affect turbidity and cloud formation, and these processes may augment or counteract the warming effects of the greenhouse gases. Some mechanisms of this kind are discussed under "Feedbacks in the Climate System," below.

INCREASES IN EMISSION OF GREENHOUSE GASES DUE TO HUMAN ACTIVITIES

The greenhouse gases of greatest concern in the context of this paper are CO_2, CH_4, N_2O, and the CFCs. These are increasing in their atmospheric concentrations, at rates that could lead to significant warming of the atmosphere in the coming century. The 1990 IPCC report provides information on their preindustrial atmospheric concentrations, present day concentrations and current rates of increase (see table 1).

Carbon Dioxide

Carbon dioxide had risen in concentration from a preindustrial concentration of near 280 parts per million by volume (ppmv) to 353 ppmv in 1990 and continues to increase at about 0.5%, or 1.8 ppmv per year. The IPCC estimates that CO_2 contributed 55% to the increase in radiative forcing caused by all the greenhouse gases between 1980 and 1990.

Annual emissions of CO_2 are estimated at 5.9 gigatonnes of carbon (GtC) from fossil fuel use and between 0.4 and 2.6 GtC from deforestation. But the 1.8 ppmv/yr concentration increase in the atmosphere is equivalent to only about 3.4 GtC of the 6.3 to 8.5 GtC emitted. A major uncertainty about CO_2 is the exact disposition of the excess carbon. Clearly, however, the missing CO_2 must be distributed between the land and the oceans—the oceans having been thought until recently to be the major sink. Tans et al. (1990), on the basis of observed air-sea CO_2 gradients, concluded that oceanic uptake of atmospheric CO_2 is smaller and that more carbon has been and is being captured by vegetation on land than had previously been believed. However, Watson et al. (1991), on the basis of measurements of atmospheric and surface water CO_2 concentrations in the North Atlantic, conclude that the uncertainties about oceanic uptake of CO_2 are still large enough to make the Tans et al. conclusions premature. The disposition of the "missing CO_2" is a major uncertainty in carbon balance research—an uncertainty that makes prediction of future atmospheric CO_2 concentrations very difficult.

Methane

Methane appears to be the greenhouse gas next in importance to CO2 in radiative forcing. Its concentration has more than doubled from a preindustrial 0.8 ppmv to 1.72 ppmv in

Table 1. Findings of the IPCC (1990) with respect to major radiatively active trace gases[a]

	Carbon dioxide	Methane	Nitrous oxide	CFC11	CFC12
Atmospheric concentration[b]	ppmv	ppmv	ppbv	pptv	pptv
Preindustrial (1750-1800)	280	0.8	288	0	0
Present day (1990)	353	1.72	310	280	484
Current rate of change:	1.8	0.015	0.8	9.5	17.0
per year	(0.5%)	(0.9%)	(0.25%)	(4.0%)	(4.0%)
Atmospheric lifetime (years)	50-2000[c]	10	150	65	130
Reduction in human-made emissionsrequired to stabilize concentrationsat present-day levels	>60%	15-20%	70-80%	70-75%	75-85%
Warming potential of 1 kg of gas emitted today after 20 years (relative to CO2)	1	63	270	4,500	7,100
Contribution to radiative forcing from 1980 to 1990 (%)	55	15	6	——— (17%) ———	

Notes:
[a] Uncertainty due to lack of knowledge concerning dynamics of oceanic and biospheric absorption.
[b] ppmv, ppbv, and pptv refer, respectively, to parts per million, billion, and trillion by volume.
[c] Compiled by Lemon et al. (1992) from a number of tables and figures in the Policymakers Summary and Part 1 of IPCC (1990).

1990, and is increasing now at a rate of about 0.9% per year.[2] The IPCC (1990) estimates CH_4 contributions to radiative forcing in the last decade at 15%. Current understanding suggests that a given quantity of CH_4 emitted into the atmosphere today will, after 20 years, be about 60 times more effective in warming the atmosphere than an equal mass of CO_2 emitted today (see table 1).

Sources of CH_4 are both industrial and biogenic. Industrial sources include leakage from natural gas transmission and distribution systems and escape from coal mines and oil and natural gas wells. Biogenic emissions are the result of anaerobic decomposition of organic matter in natural wetlands, rice paddies, and landfills. CH_4 is also produced as a by-product of animal digestion, particularly in ruminants, and by termites and other insects.

According to Crutzen (1991), the total annual source of CH_4 is 550±105 Tg/yr (Tg refers to teragrams; 1 Tg = 10^{12} grams = 10^6 metric tons). About 100±20 Tg/yr come from fossil fuels and CH_4 hydrates.[3] If so, another 405 Tg/yr or so must come from biogenic sources. The IPCC (1990) scientific assessment estimates the range of current CH_4 emissions (all in units of Tg/yr) as follows: 100 to 200 from natural wetlands; 25 to 170 from rice paddies; 65 to 100 by enteric fermentation (animals); 10 to 100 by termites; 20 to 70 from landfills; 20 to 80 from biomass burning; 1 to 25 from fresh waters; 5 to 20 from oceans.

This wide range of estimates indicates great uncertainty in our quantitative knowledge of biogenic CH_4 emissions. In a study recently completed at Resources for the Future, Lemon, Katz, and Rosenberg (1992) have assessed the reasons for this situation. Emissions estimates are uncertain because of problems associated with (1) the measurement or estimation of gaseous fluxes, (2) the measurement or estimation of areas of the emitting ecosystems or populations of the emitting organisms, or (3) systematic problems in integrating from a very limited observational base to the global scale.

Uncertainties of flux measurements stem from shortcomings in the instrumentation employed and from inadequate spatial and temporal sampling. Flux of CH4 is controlled by soil temperature, moisture content, aeration, acidity, and other rate-limiting factors. Lack of decisive correlations between these factors and actual emissions makes generalizations from the very limited numbers of measurements very difficult. It is also now clear that soils provide a strong sink for CH_4 (e.g., Whalen et al., 1991).

Uncertainties in knowledge of the "area factor" stem from inadequate definition of subecosystem variations that affect emissions and from difficulties in accurately mapping land areas according to those distinctions. In the case of CH_4-emitting domestic animals, difficulties arise in assembling accurate statistics on population and nutritional status, especially (but not only) in developing countries. Difficulties are still greater in counting wild ruminants and in establishing the populations, by kind, of CH_4-emitting insects.

Nitrous Oxide

Nitrous oxide has increased from its preindustrial atmospheric concentration of 288 parts per billion by volume (ppbv) to 310 ppbv in 1990. Its current rate of increase is 0.25% per year. Even though it is accumulating more slowly than CO_2 or CH_4, its effective warming potential per unit mass emitted today will, according to IPCC (1990), after 20 years, be about 270 times greater than that of CO_2.

It now appears that N_2O emissions are primarily biogenic in origin, previous estimates of large industrial sources having been due to measurement error. The biogenic emissions stem from biomass burning (Cofer et al., 1991); natural, disturbed, and cultivated soils (Bouwman, 1990; Cicerone, 1989; Mosier and Hutchinson, 1981); and, possibly, from nitrogen-containing groundwaters (Ronen et al., 1988).

According to the IPCC's 1990 estimates, combustion processes account for only 0.1 to 0.3 Tg N/yr out of a possible total of 4.4 to 10.5 Tg N/yr. Biogenic sources account for the remainder. Forest soils produce 2.9 to 5.2 Tg N/yr; biomass burning, 0.02 to 0.2; fertilizer (including fertilizer nitrogen in groundwater), 0.01 to 2.2; and oceans, 1.4 to 2.6 Tg N/yr. Ronen et al. (1988) estimate emissions from groundwater to range from 0.8 to 1.7 Tg of N_2O.

Uncertainty of the estimates of biogenic N_2O emissions stems from the same causes as with CH_4: inadequate measurements, uncertain estimates of the areas of ecosystems involved, and difficulties of integrating up from a very limited data base to the global scale.

The CFCs

CFCs 11 and 12 are found in the atmosphere in concentrations of 280 and 484 parts per trillion by volume (pptv), respectively, (1990 values). Both are increasing at rates of approximately 4.0% per year. The CFCs are strongly absorptive of longwave infrared radiation. Hence they are strong greenhouse gases, and IPCC (1990) attributes 17% of the 1980-90 radiative forcing to the CFCs. Also, the CFCs have been shown quite conclusively to be the source of chlorine atoms that catalyze the photolytic destruction of O_3 in the atmosphere (Watson et al., 1988).

The CFCs illustrate well the rapidity with which new scientific findings relating to greenhouse warming replace not-so-old ones. It now appears (Ramaswamy et al., 1992; Allbritton[4]) that the O_3 losses attributable to the CFCs could actually lead to atmospheric cooling rather than warming. The ozone losses cause a local cooling in the lower stratosphere, which then radiates less heat toward the surface, leading to a cooler lower atmosphere/surface system.

This indirect cooling tendency may offset the direct warming due to the "greenhouse" behavior of the CFCs and other gases. Thus the CFCs appear to be dropping in importance as greenhouse gases (but not as O_3 depleters), and other greenhouse gases must account for larger percentages of the total greenhouse radiative forcing than previously assumed. The net effect of the CFCs as greenhouse gases may, in fact, be near zero. It also appears that the total greenhouse forcing may have been overestimated by perhaps 20% because of the net forcing previously attributed to the CFCs.

CLIMATE CONSEQUENCES OF GREENHOUSE WARMING

Information was presented in the previous section on how the concentration of greenhouse gases in the atmosphere is growing. What might be the climatic consequences of these changes? To learn how things work, scientists pose hypotheses that they then test by experimentation. However, the global climate system is complex, and our ability to perform controlled experiments on it is severely limited (although humankind may well be performing uncontrolled experiments on it right now). Theoretical models that consider all components of the climate system—atmosphere, oceans, cryosphere, land masses, and biosphere—provide the best means of predicting how natural and man-made phenomena can alter the workings of the climate system (Lawrence Livermore National Laboratory [LLNL], 1990).

The General Circulation Model

The general circulation model (GCM) has become the primary tool for calculating or "experimenting" with the atmosphere to determine how changes such as increasing greenhouse gas concentrations might change climate.[5] I have seen no better explanation of the GCM than that of LLNL (1990):

> GCMs divide the global atmosphere into tens of thousands of discrete boxes and use the dynamical equations of motion, energy and mass to predict the changes in winds, pressure, and water vapor mixing ratio (humidity). The vertical domain of GCMs typically extends from the earth's surface to about 35 km; this distance is divided into from two to twenty computational levels. The horizontal domain covers the globe with grid cells, each of which is several hundreds of kilometers on a side.

What Do the GCMs Predict?

A survey of models done by IPCC (1990) indicates "with high confidence" that an equilibrium climate change due to a doubling of CO_2 or its radiative equivalent due to all the greenhouse gases will warm the lower atmosphere (troposphere) and cool the stratosphere, and that global average tropospheric warming will, by the end of the 21st century, range between 1.5°C and 4.5 C, with 2.5°C as the "best guess."[6] Global average precipitation will increase, as will evapotranspiration, and the more the warming, the greater the increase for both. Global average precipitation and evapotranspiration are estimated to increase by 3% to 15%.

The changes described above will not be uniformly distributed around the globe. IPCC indicates, with a lesser degree of certainty than attributed to global mean temperature, that high latitudes will warm more than the global average in winter but less in summer, and that surface warming and its seasonal variation will be least in the tropics. Precipitation is projected to increase in the high latitudes throughout the year and in the mid-latitudes in winter. Zonal mean precipitation will increase in the tropics, although there will be areas of decrease as well. Models differ considerably in the shifts of tropical rainbands that they predict. Little change is expected in precipitation in the subtropical arid regions. As a consequence of the above, soil moisture should increase in the high latitudes and decrease in the mid-latitudes of Northern Hemisphere continents in summer. The areas of sea ice and seasonal snow cover should diminish.

How Well Do the GCMs Agree?

None of the 20 or so GCMs in current use can produce perfectly accurate representations of even current climatic conditions. While all are of basically the same design and derive in many ways from one another, the changes in climate they project for the future as the result of any particular change in atmospheric conditions (say, a doubling of their preindustrial CO_2 concentration or its radiative equivalent in all the greenhouse gases) differ in their estimates of global mean climate change and, particularly, in the regional distributions of change. It is impossible to know now which of the 20 or so extant models is most accurate and which will prove the best predictor of climatic conditions when the CO_2 effective doubling or any other change in total radiative forcing actually occurs. But it is possible, through intercomparisons, to determine why the models differ and through this process find ways to improve their reliability. A number of efforts to accomplish such intercomparisons are already under way (e.g., LLNL, 1990; U.S. Department of Energy [DOE], 1990; Cess et al., 1991).

There are a number of reasons for the different results that emanate from current models and the general concern for their reliability. Grid cells in the current models are too large to allow important processes with global implications, such as those controlling cloud size and amount, to be captured. It is also generally agreed that the GCMs require improved parameterization of biologic processes (plant growth, vegetation cover, and roughness, for example), improved treatment of cloud physics and effective coupling of atmospheric and oceanic dynamics, and better understanding and parameterization of feedback processes if significant improvements are to be made. Efforts are under way in many modeling groups to accomplish these improvements (e.g., Wilson et al., 1987; Dickinson and Hanson, 1984; Dickinson et al., 1987a, b; Schlesinger and Jiang, 1988; Hansen et al., 1985; Washington and Meehl, 1984, 1989).[7]

Feedbacks in the Climate System

The climate system is not yet fully understood. Hence, the predictions of the GCMs involve a number of important uncertainties related to the interactions of atmosphere, oceans, cryosphere, and biosphere. Interactive feedbacks can either amplify or dampen the climate response to additional greenhouse gases.

Following are some possible climatic feedbacks to greenhouse warming:

1. Because of increased evaporation, water vapor increases in the atmosphere as it warms; water vapor is a strong greenhouse gas: positive feedback.
2. Snow and ice melt as the atmosphere warms; snow and ice have high albedo; less solar radiation is reflected to space, more is absorbed: positive feedback.
3. The oceans warm; CO_2 dissolved in the upper layers is released to the atmosphere, further increasing its CO_2 concentration and greenhouse forcing: positive feedback.
4. Warming causes melting of permafrost in the arctic, exposing organic materials to respiratory decay and release of CO_2 into the atmosphere: positive feedback.
5. Warming in high latitudes causes the release of CH_4 trapped in hydrates: positive feedback.

6. But warming dries the soil; dry soils are a sink for CH_4 and can lower its atmospheric concentration: negative feedback.

7. Warming and increased water vapor in the atmosphere lead to changes in cloud amount, cloud altitude, and cloud water content: both positive and negative feedbacks are possible.

Space does not permit review of each of these feedback mechanisms and the many others, not listed here, that are possible. But the conflicting views concerning a few of the feedbacks demonstrate why the extent of future radiative forcings and the climatic consequences predicted by the GCMs remain so uncertain.

Water Vapor Feedback Increases Radiative Forcing

Lindzen (1990a, b) has stated that the warming potential of greenhouse gas increases is overestimated by the GCMs because the models distribute the increased atmospheric water vapor (increased because warming leads to increased evaporation) uniformly throughout the atmosphere. Water vapor is, of course, a strong greenhouse gas itself. Lindzen proposes a mechanism by which convective processes set in motion by greenhouse warming could dry the upper troposphere, with the result that warming would be much smaller than the GCMs predict. Betts (1990), among others, disputes Lindzen's proposed negative feedback mechanism on theoretical grounds. Shine and Sinha (1991) show in a modeling experiment that the atmospheric radiation budget is more sensitive to changes in vapor content of the lower troposphere than to the same percentage changes in the upper troposphere. Thus, even if Lindzen's hypothesis is correct, there would be little effect on overall warming. Kellogg (1991) cites evidence of actual warming in the troposphere over the tropics in the last few decades and increased water vapor content above the equatorial Pacific as evidence against Lindzen's hypothesis. However, it is not yet clear whether these observations are indicative of global trends and are "greenhouse signals" as such, or merely "noise" in the variable climatic system.

Snow and Ice Melting Reduce Albedo

Even this fairly intuitive outcome of global warming may be more complicated than previously thought. Miller and de Vernal (1992) find geological data supporting the idea that greenhouse warming, expected to be most pronounced in the arctic and in winter, if coupled with decreasing summer insolation (due to the natural orbital cycles), could lead to a situation where snow deposition exceeds melting at high northern latitudes and hence to ice-sheet growth and the inception of glaciation.

Ledley (1991) proposes another interesting hypothesis. If sea ice is transported equatorward, the sea-ice pack thins. More open ocean occurs, and the poles are ice-free for longer periods. Since show and ice insulate the unfrozen water from the atmosphere, more energy is transferred from the ocean to the atmosphere near the poles. The movement of the ice leads to a warming at all affected latitudes greater than the cooling caused in the latitudes into which the ice moves.

Cloud Feedbacks

Clouds play an important role in regulating earth's temperature. Clouds contribute to the greenhouse warming of the earth but also act to cool it by reflecting solar radiation to space. The net effect of clouds today is a 13 W m^{-2} radiative cooling (IPCC, 1990), or about 4% of the mean radiative flux at the top of the atmosphere.

If global warming decreases cloud amount, which the GCMs generally predict will happen, earth cools more effectively as thermal radiation escapes more easily to space; however, reflection of incoming solar radiation to space also diminishes, and the climate system absorbs more solar radiation, which would warm the system. If a cloud layer is displaced, say, to a higher elevation and lower temperature, less thermal radiation would be emitted to space. This would result in a net increase in energy retained in the atmosphere. Displacement to a lower elevation would have the opposite effect.If cloud water content increases, clouds would become brighter, with a resultant negative feedback, but positive feedback effects are also possible, depending on elevation of the cloud, since the thicker cloud is a stronger absorber of infrared radiation.

Dimethyl sulfide (DMS) is a gas released by certain species of marine phytoplankton and bacteria. The gas, oxidized in the atmosphere, forms compounds that act as cloud condensation nuclei (CCNs) (Charlson et al., 1987). It is hypothesized that, under warm conditions, the marine organisms would produce more DMS, leading to increases in cloudiness—particularly low stratiform clouds—over the oceans.

The linkage of DMS and atmospheric sulfates appears to have been demonstrated by Ayers et al. (1991). A more tentative linkage between DMS and CCN formation has also been observed by Ayers and Gras (1991). The DMS/CCN linkage is thoroughly described in Charlson et al. (1992).

The complexity of the cloud feedback question is further emphasized by the results of an intercomparison of 17 GCMs (Cess et al., 1991) in which the climatic response to different parameterizations of cloud conditions was varied over a very wide range. Physically consistent positive and negative changes in global mean warming were found.

What Does the Record Show about Global Warming?

According to IPCC (1990), global mean temperature has increased by 0.3°C to 0.6°C over the past century—an increase consistent with expected greenhouse warming, but not in

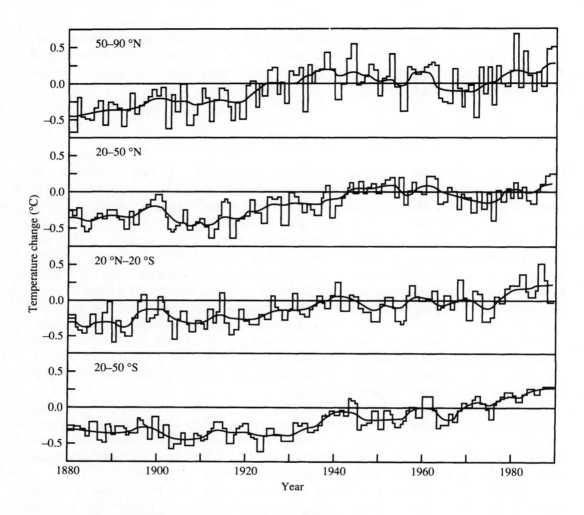

Figure 2. Observed variations in annual-mean tempature for various latitude bands. The temperatures used are air tempature data over land areas and sea-surface tempaerature data for the oceans. The smooth curves are filtered values designed to show decadal and longer time-scale trends more clearly (Source: Figure 8.3 in IPCC, 1990).

itself evidence of this warming. Figure 2, from IPCC (1990, p. 250, fig. 8.3), shows observed variations in annual mean temperature in various latitudinal belts. The figures are constructed using air temperature data over land and sea surface temperature for the oceans. Over the past 100 years, high northern latitudes have warmed slightly more than the global mean, but only in winter and spring. Except in recent years there has been little evidence of warming in these latitudes. In the 20°N to 50°N belt, temperatures rose until the 1940s, leveled out, decreased in the 1960s, and are rising again. Trends in the Southern

Hemisphere (where the GCMs generally predict lesser warming) more clearly demonstrate a warming trend during the past century.

The observed warming, if due entirely to an anthropogenic greenhouse effect, would be near the lower end of the range predicted by GCMs for the radiative forcing of greenhouse gases added to the atmosphere in the past 100 years. If natural variability accounts for some of the observed warming, then the climate sensitivity would be still lower. IPCC (1990) suggests the possibility that larger greenhouse warming could have been offset in part by natural variability (on the cooling side) and by other factors, also perhaps anthropogenic.

Interestingly, studies of the climate record for the 48 contiguous states (Karl et al., 1991; Hanson et al., 1989) show no long-term trend in annual mean temperatures. Although the 48 states comprise a very small fraction of the earth's total surface, this region has one of the most dense and reliable weather networks. Recently weather records have revealed that while mean temperatures have not risen, nighttime temperatures have risen slightly (Karl et al., 1991). Observations in the Soviet Union and China confirm these results. Such an effect is consistent with increased atmospheric water vapor content and/or increased cloudiness, both of which reduce the escape of thermal radiation to space.

It now appears possible that increases in global haze could be responsible for holding back the warming that GCMs indicate should already have been observed and for increasing nighttime temperatures. Sulfur emitted from smokestacks is oxidized to sulfuric acid droplets, which are capable of reflecting sunlight back to space. Shading by a sulfate haze could be counteracting almost half of the global temperature rise that would have occurred as the result of increases in concentrations of all the greenhouse gases added to the atmosphere since preindustrial times (Kerr, 1992a; Charlson et al., 1992). Another possible shading factor, perhaps as important as the sulfate aerosols, has recently been identified by Penner et al. (1992). Smoke particles from biomass burning reflect solar radiation directly and also act as CCNs, increasing cloud reflectivity. Together these smoke effects could cause a cooling effect of as much as 2 Wm^{-2}. The authors think it possible that anthropogenic increases of smoke emission may have helped to weaken the net greenhouse warming from greenhouse gas emissions.

Other evidence of global warming is in short supply. Lachenbruch and Marshall (1986) observed a recession of permafrost in boreholes on and near the north slope of the Alaskan Arctic that would be consistent with long-term warming of 2°C to 4°C. Michaels et al. (1988), however, contend that the temperature records for the region since 1910 are too noisy to support any temperature trend and, further, that any warming determined for the permafrost record would have occurred before the major emissions of greenhouse gases began.

Sea-level change is not treated as a major issue in this paper. However, changes in sea level are expected to be among the most important consequences of global warming. The sea-level record can also be used as an indicator of whether the climate is already warming. IPCC (1990) states that increasing greenhouse gas concentrations have caused global mean sea level to rise, partly because of oceanic thermal expansion and partly because of the melting of land-based ice masses.

The evidence for sea-level rise in the past century is strong, although the estimates vary by a factor of at least two. Observed and calculated rates of global sea-level change range from about 0.5 to 2.4 millimeters per year (mm/yr), with a standard deviation of 0.9 mm/yr. Gornitz et al. (1982) show an almost linear rate of increase of 1.2 mm/yr, while Barnett (1985) suggests 1.15 mm/yr, but with a steeper rate of 1.7 mm/yr in the period from 1910 to 1980. The consensus view of IPCC is that the increase in mean global sea level is attributable to thermal expansion of the oceans and increased melting of mountain glaciers and the edges of the Greenland ice sheet.

DIRECT EFFECTS OF CO_2 ON PLANTS

Fertilization and Water Use

We turn now to a different but closely related subject. It is a well-known and demonstrable fact that plants, when exposed to increased concentrations of CO_2, respond with an increased rate of photosynthesis.[8] Such increases in photosynthesis normally lead to larger and more vigorous plants and to higher yields of total dry matter (roots, shoots, leaves) and often of fruits, grains, etc. The behavior described here is demonstrated particularly by plants of the C_3 category, which includes most species—small grains, legumes, root crops, cool season grasses, and trees. Another category of plants, the C_4, or tropical grasses, such as corn, sorghum, millet, and sugar cane, are naturally more efficient photosynthesizers than the C_3 plants. They too respond, but less markedly, to increases in atmospheric CO_2.

However, the C_4 plants show another interesting response to increased CO_2 in the atmosphere: Their consumption of water by transpiration is reduced because of partial closure of the leaf stomata (pores) induced by high CO_2 concentration. The effect is also demonstrated by the C_3 plants. The reduction in transpiration is not accompanied by any significant loss in photosynthesis.

A recent experiment by Chaudhuri et al. (1990) illustrates the kinds of results most often obtained in agronomic experiments with elevated atmospheric CO_2. The authors grew winter wheat over three growing seasons at 340 (ambient), 485, 660, and 925 ppmv of CO_2 at both high and low levels of water supply. Yields of wheat increased in both water regimes and at all levels of CO_2 above the ambient. In this experiment, water requirement was reduced by elevated CO_2, with the greatest reduction in almost all cases with the first increment of added CO_2. Water use efficiency (crop production per unit of water consumed) was increased because of reduced transpiration and increased photosynthesis.

If the findings from laboratory, greenhouse, and open chambers upon which the above statements are based can be extrapolated to the open field, we may expect important benefits to agriculture from the increasing concentration of CO_2 in the atmosphere—increased photosynthesis in many important species and decreased water consumption in most species. This direct "fertilization" effect of CO_2 should offset detrimental climatic changes to some extent. Unmanaged ecosystems, where stability is highly valued, may also be subjected to important changes. But in cases where the climatic changes are clearly beneficial and increase the potential for production—for example, through increased warmth in the northern parts of the temperate zone or increased soil moisture in the semiarid regions—the fertilization effect increases the likelihood that the greater production potential can be realized.

In the 1970s and early 1980s, debate was vigorous (e.g., Lemon, 1976; Rosenberg, 1981) on the question of whether these laboratory-demonstrable CO_2 effects on photosynthesis and transpiration do now or will occur in the future in the field, where temperature, moisture, and nutrients are the factors that normally limit plant productivity. Since then, further laboratory, controlled environment, and open-top chamber experiments have shown that CO_2 enrichment of the atmosphere actually reduces the impacts of moisture and salinity stress on plants (Rosenberg et al., 1990). A summary by Allen et al. (1990) shows that high-temperature stress is also alleviated by CO_2 fertilization. The effects on nutrient stress remain less clear at this writing.

It is possible, although certainly not yet proven, that some portion of the overall increase in crop yields that has occurred in the last 100 years has been due to the increase in atmospheric concentration of CO_2. Direct field evidence of this "CO_2 fertilization effect" eludes us, since it is not yet possible to identify a CO_2 signal in the noisy records of annual crop yields. However, increased capture of CO_2 by the terrestrial biosphere (including agricultural ecosystems) ought to be occurring as a result of rising CO_2 levels in the atmosphere. Recent findings are consistent with this view. For example, Kohlmaier et al. (1987) interpret the increasing amplitude of the CO_2 concentration wave at Mauna Loa as suggesting an increasing global biomass. This could be due, at least in part, to CO_2 stimulation of plant growth and would be consistent with the assertion by Tans et al. (1990) that more of the "missing carbon" has been and is being captured by vegetation on land, and less by the oceans.

CO_2 enrichment of the atmosphere could also lead to some troublesome effects. Because of the special benefit that C_3 plants derive from elevated CO_2, C_3 weeds may become more competitive in fields of C_4 plants. Additionally, the C:N ratio increases in leaves of plants fertilized with CO_2. Some short-term studies show that because of this change, insects must consume more vegetation to satisfy their nutritional needs (Lincoln et al., 1984), implying greater insect damage. Later studies suggest more complicated outcomes, however. Over longer periods, the population of insects feeding on plants stimulated by higher CO_2 would likely decline in response to the diminished proportion of nitrogen in the plant tissues. As pest populations decline, so too would the populations of their predators (Fajer et al., 1989). Although changes in the provenance of weeds, insects, and diseases are likely to accompany changes in climate, the long-term ecological changes that might follow from increasing CO_2 concentration in the atmosphere are difficult to predict (Bazzaz and Fajer, 1992). However, the impacts on agriculture of changes in weed competition and insect activity would probably be less profound than they would be in unmanaged ecosystems (Rosenberg, 1992).

Results of Field Experiments

The foregoing information on plant responses to CO_2 enrichment has been derived almost entirely from agronomic and horticultural experiments under controlled environment conditions or in a few cases with chambers enclosing small clusters of plants in a crop field. Only a very few studies have been conducted in natural ecosystems under ambient conditions, and most of those, too, have involved the use of chambers enclosing clumps of vegetation. In most of the recent studies, chambers are opened at the top so that artificiality of the chamber environment can be minimized.

Results of two studies in widely contrasting unmanaged ecosystems are particularly interesting. Tissue and Oechel (1987) studied the response of tundra vegetation in Alaska to CO_2 enrichment over a number of years. Despite an initial positive growth response to CO_2 enrichment in one type of tussock grass, overall the species studied grew no better than plants exposed to unenriched air. Low temperatures and photosynthate accumulation (lack of capacity of the plant to store increased photosynthate) may be responsible for these results.

On the other hand, Drake (1989, and personal communication[9]) has found that marsh vegetation enclosed in open-top chambers on the shore of the Chesapeake Bay responds positively to CO_2 fertilization. This response continues after five years of observation and is greater for *Scirpus olneyi* (a C_3 plant) than for *Spartina patens* (a C_4 plant). The reasons for the different outcomes of the Alaska tundra and Chesapeake marsh experiments are not yet clear, although differences in climate are probably involved. There is evidence that response to CO_2 enrichment is limited by low temperature (Allen et al., 1990; Idso et al., 1987).

Because of the many complicating factors induced by the artificiality of experimental environments, it will be some time before we have definitive answers on whether CO_2 fertilization will actually affect photosynthesis in totally natural open-air environments. Attempts to measure open-air responses began in the 1970s in experiments where CO_2 was released directly into crop fields (Harper et al., 1973a, b for cotton; Allen et al., 1974 for corn). These experiments failed to provide definitive results, primarily because of the difficulty of maintaining elevated CO_2 concentrations in the air surrounding plants in the face of normal atmospheric turbulence, which tended to remove it rapidly.

Since then technological advances have made open-air CO_2 enrichment research possible. In a program called FACE (Free-Air Carbon Dioxide Enrichment) sponsored by the U.S. Department of Energy and conducted by the U.S. Department of Agriculture, equipment has been developed to maintain an elevated level of CO_2 in the air within and above crops for entire growing seasons (Hendrey and Kimball, 1990). The FACE system was originally tested in a Mississippi cotton field using industrial by-product CO_2 and has since been used in a field near Phoenix, Arizona. Cotton crops were grown for two years with CO_2 held at a concentration of about 550 ppmv. Cotton growth and yield has been about 40% greater with CO_2 enrichment. Seasonal water use may not have been affected, although use appears to have been greater early in the season because of more rapid crop growth, while the stomatal closure effect reduced use later in the season. Water use efficiency improved in direct proportion to the increase in yield.[10]

ON CONTROL OF GREENHOUSE GAS EMISSIONS

We know something now of the causes of greenhouse gas accumulation in the atmosphere and something of the possible effects of these gases on climate and on plant growth. Prudence suggests—not knowing what the changes will be, but recognizing the possibility that they may be severe—that greenhouse warming should be slowed or perhaps even stopped. To do so will require sharp reductions in the emissions of greenhouse gases into the atmosphere. Much has been written on how these reductions may be achieved, and various options have been debated widely. It is beyond the scope of this paper to review control issues in general. However, some insights emerging from recent scientific findings reported in this paper deserve mention here.

Since the CFCs may be less important in greenhouse forcing than previously thought, CH_4, N_2O, and CO_2 appear to be relatively more important. Controlling CO_2 emissions will require major reductions in fossil fuel use and in the rate of tropical deforestation—apparently the major biogenic source of the increasing atmospheric concentration of CO_2. Afforestation and reforestation have been proposed as means of increasing the terrestrial sink strength for CO_2, thereby diminishing the rate at which CO_2 accumulates in the atmosphere (e.g., Sedjo and Solomon, 1989). The topic of controlling CO_2 emissions—both industrial and biogenic—is covered in numerous comprehensive reports (e.g., IPCC, 1990; NAS, 1991) and is beyond the scope of this paper. So, too, is the matter of reducing industrial emissions of CH_4 and N_2O. A few remarks on reducing biogenic emissions of the latter two gases are presented here.

Biogenic Methane

Of the biogenic sources of *methane,* only a few are subject to control unless, of course, decisions are made to drain the world's natural wetlands and to eliminate the world's termite populations. Such "solutions" are neither practical nor advisable; indeed they are not even within the realm of the possible. Emissions from fresh waters and oceans are probably beyond control. Thus, control, if any, will have to be applied through reductions in biomass burning, changes in rice paddy culture, and management of ruminant animal populations and landfills.

According to a report of the U.S. Environmental Protection Agency (Lashof and Tirpak, 1990), a number of possible control strategies are under study. In rice culture, for example, shifts from paddy to upland rice, changes in manuring practices to diminish the quantities of organic materials incorporated in paddy soils, changes in crop residue disposal practices, and changes in rice varieties to diminish the quantities of straw to be disposed of have all been proposed as means of diminishing CH_4 emissions. Increased use of nitrogenous fertilizers and improvements in the efficiency of fertilizer use can also decrease the need for organic manures.

Emissions from ruminant animals may be subject to some control through management of nutrition and through the use of CH_4-suppressing feed supplements. One proposal is to increase the efficiency of milk and meat production through the use of a genetically engineered substance—BGH (bovine growth hormone). By this means, current needs for milk and meat might be met with fewer animals and, therefore, with lesser emissions of CH_4. Systems to capture CH_4 emitted in landfills are already in use.

Biogenic Nitrous Oxide

Methods have also been proposed to reduce nitrous oxide emissions from fertilized soils (Lashof and Tirpak, 1990). Since fertilizer type has an effect on rate of emissions, a shift from anhydrous ammonia to other forms of fertilizer nitrogen offers some opportunity for reducing nitrification. Certain crops, like corn, require much larger quantities of nitrogen than do others, like wheat and especially legumes such as soybeans. Thus shifts in the mix of crops grown can have some impact on emissions. Changes in the depth of fertilizer placement and timing of applications are also known to affect the rates of emission of N_2O. Additives can be used with nitrogen fertilizers to inhibit nitrification and to reduce the rate of release. The use of nitrification inhibitors may be economical in the future, and some decreases in N_2O emissions may result.

Outlook

The outlook, then, for reducing greenhouse gas emissions other than CO_2 appears to be as follows: the large reductions in CFC emissions (70% to 85%) required to stabilize their concentrations in the atmosphere (IPCC, 1990) will probably be achieved as a result of compliance with the provisions of the Montreal Protocol and its amendments (United Nations Environment Programme [UNEP], 1987), although it now appears that this control may have little effect on greenhouse warming.

Reducing the industrial emissions of CH_4 and N_2O should be technologically feasible, now requiring only the encouragement that a proper mix of economic incentives and penalties can provide. Reducing the biogenic emissions of these gases will be more difficult, however. Certain important sources of CH_4, such as wetlands and termites, are (or should be) left alone for sound ecological reasons. The controllable biogenic sources—paddy rice and cattle for CH_4 and nitrogenous fertilizers for N_2O—present different problems. The world's acreage of rice paddies and the numbers of its ruminant animals are large. But individual holdings and herds are small. This means that great numbers of farmers and herders in developing and developed countries alike will need to acquire the skills required to manage the lands or the animals in ways that diminish emissions. For CH_4 and N_2O emissions reductions to occur on a large enough scale to make a difference will require major investments in extension education and the economic incentives to encourage the adoption of appropriate methods.

SUMMARY

Driven by a variety of natural causes, climate is continually changing. Now, however, human activities, particularly land use change and the emissions of greenhouse gases into the atmosphere, appear likely to induce other changes in climate, and at an unprecedented rate. The most likely change is a general warming of the lower layers of the atmosphere due to enhancement of the planet's natural greenhouse effect. This change will be forced by the rising concentrations in the atmosphere of CO_2, CH_4, N_2O, and the CFCs. But predicting when, where, and by how much not only temperature but also precipitation, windiness, cloudiness, soil moisture, etc., will change and what the consequences of these changes might be is extremely difficult. Current projections made with GCMs encompass great uncertainties because of the complexity of climate dynamics and the many feedback processes that greenhouse warming might invoke.

Of the greenhouse gases now increasing in atmospheric concentration because of human activity, CO_2 is the most important from the point of view of radiative forcing. Increased atmospheric concentration of CO_2 has also been shown to increase plant growth and reduce water use in agricultural crops, and it likely does so in other terrestrial ecosystems as well. While supporting evidence is difficult to obtain, it seems likely that the CO_2 fertilization effect is already being expressed and is responsible for the sequestration in terrestrial ecosystems of a large portion of the CO_2 emitted into the atmosphere by fossil fuel combustion and deforestation. These direct (nonclimatic) effects of CO_2 will be beneficial to agriculture and forestry. It is not at all clear, however, that the changes in the dynamics, species distribution, or provenance of unmanaged ecosystems because of CO_2 fertilization would be beneficial.

On balance, the possible harm that climatic change induced by greenhouse warming might do probably outweighs the benefits of a CO_2 fertilization effect. Most scientists would agree, therefore, that all reasonable strategies should be employed to limit or reduce anthropogenic emissions of the greenhouse gases to avoid climate change. Recent findings suggest that the CFCs play a much smaller role in radiative forcing than we thought only a year ago. Control of CH_4 and N_2O from biogenic sources will be difficult. Thus, the burden of greenhouse-effect avoidance remains squarely on those industries and activities that emit CO_2 into the atmosphere through consumption of fossil fuel and through tropical deforestation.

ACKNOWLEDGMENTS

I am indebted to Dr. Mack McFarland of E. I. du Pont de Nemours & Company for helpful criticism of an earlier draft and to Daniel Balzer and Angela Blake for bibliographic and stenographic assistance in the preparation of this paper.

NOTES

1. Large volcanoes can affect global temperature in opposite ways from, but to an extent similar to, the anthropogenic greenhouse gas emissions discussed later. Eruptions such as those of Mount Agung in Bali in 1963 (Hansen et al., 1978), El Chichon in Mexico (Kerr, 1983), and Mount Pinatubo (Kerr, 1992a) injected large quantities of particles and gases into the stratosphere, increasing planetary albedo and cooling the atmosphere by a degree or so for a few years.

2. Vaghjiani and Ravishankara (1991) have found that the rate coefficient for the reaction of hydroxyl radicals with atmospheric CH_4 is smaller than had been assumed. This finding indicates an approximately 25% longer lifetime for CH_4 and a correspondingly smaller flux (by about 100 Tg CH_4/yr) than previously calculated.

3. Gas hydrates may be present under certain temperature-pressure conditions found in permafrost regions and in sea sediments of the outer continental margins. These hydrates are solids composed of water molecules with large amounts of gas (mainly CH_4) trapped in them (Svensson et al., 1991).

4. Personal communication, D. Allbritton, NOAA Aeronomy Laboratory, Boulder, CO, January 1992.

5. Lamb (1986) lists alternatives to the use of GCMs, such as analogs from the climate record and paleoclimatic reconstructions, and discusses strengths and weaknesses of these as well as the GCM approach.

6. This range was estimated prior to the recent evidence of a change in net radiative forcing by CFCs 11 and 12.

7. To assist in these efforts, a program to quantify uncertainties in the current crop of climate simulation models (MECCA, Model Evaluation Consortium for Climate Assessment), supported by EPRI and other organizations, has provided a supercomputer to the National Center for Atmospheric Research to be dedicated to climate studies.

8. Although many such studies are conducted at CO_2 concentrations intended to roughly double the preindustrial concentration (600 to 660 ppmv), the photosynthetic response is also demonstrable at lower levels (Kimball et al., 1990).

9. Personal communication, B.G. Drake, Smithsonian Institution, Edgewater, MD, May 1991.

10. Personal communication, Dr. Bruce Kimball, USDA/Water Conservation Laboratory, Phoenix, AZ, February 1992.

WORKS CITED AND GENERAL REFERENCES

Allen, L.H., Jr., R.L. Desjardins, and E.R. Lemon. 1974. "Line Source Carbon Dioxide Release. I. Field Experiment." *Agronomy Journal*, vol. 66, pp. 609-615.

Allen, S.G., S.B. Idso, B.A. Kimball, J.T. Baker, L.H. Allen, Jr., J.R. Mauney, J.W. Radin, and M.G. Anderson. 1990. *Effects of Air Temperature on Atmospheric CO_2—Plant Growth Relationships.* Report DOE/ER-0450T (TR048) (Washington, DC: U.S. Department of Energy, Office of Energy Research).

Ayers, G.P., and J.L. Gras. 1991. "Seasonal Relationship Between Cloud Condensation Nuclei and Aerosol Methanesulphonate in Marine Air." *Nature,* vol. 353, pp. 834-835.

Ayers, G.P., J.P. Ivey, and R.W. Gillett. 1991. "Coherence Between Seasonal Cycles of Dimethyl Sulphide, Methanesulphonate and Sulphate in Marine Air." *Nature,* vol. 349, pp. 404-406.

Ardanuy, Philip E., and H. Lee Kyle. 1986. "Observed Perturbations of the Earth's Radiation Budget: A Response to the El Chichon Stratospheric Aerosol Layer?" *Journal of Climate and Applied Meteorology,* vol. 25, pp. 505-516.

Barnett, T.P. 1985. "Long-Term Climatic Change in Observed Physical Properties of the Oceans." Pp. 91-107 in M.C. MacCracken and F.M. Luther, eds., *Detecting the Climatic Effects of Increasing Carbon Dioxide.* Report ER-0235 (Washington, DC: U.S. Department of Energy).

Bazzaz, F.A., and E.D. Fajer. 1992. "Plant Life in a CO2-Rich World." *Scientific American,* vol. 266, no. 1, pp. 68-74.

Bernabo, J.C., and Thompson Webb, III. 1977. "Changing Patterns in the Holocene Pollen Record of Northeastern North America: A Mapped Summary." *Quaternary Research,* vol. 8, pp. 64-96.

Betts, A.K. 1990. "Greenhouse Warming and the Tropical Water Budget." *Bulletin of the American Meteorological Society,* vol. 71, pp. 1464-1465.

Bouwman, A.F. 1990. "Exchange of Greenhouse Gases Between Terrestrial Ecosystems and the Atmosphere." Chapter 4 in A.F. Bouwman, ed., *Soils and the Greenhouse Effect* (New York: John Wiley & Sons).

Cess, R.D., and 32 others. 1991. "Interpretation of Snow-Climate Feedback as Produced by 17 General Circulation Models." *Science,* vol. 253, pp. 888-892.

Charlson, R.J., J.E. Lovelock, M.O. Andreae, and S.G. Warren. 1987. "Oceanic Plankton, Atmospheric Sulphur, Cloud Albedo and Climate." *Nature,* vol. 328, pp. 655-661.

Charlson R.J., S.E. Schwartz, J.M. Hales, R.D. Cess, J.A. Coakley, Jr., J.E. Hansen, and D.J. Hofmann. 1992. "Climate Forcing by Anthropogenic Aerosols." *Science,* vol. 255, pp. 423-430.

Chaudhuri, U.N., M.B. Kirkham, and E.T. Kanemasu. 1990. "Carbon Dioxide and Water Level Effects on Yield and Water Use of Winter Wheat." *Agronomy Journal,* vol. 82, pp. 637-641.

Cicerone, R.J. 1989. "Analysis of Sources and Sinks of Atmospheric Nitrous Oxide (N_2O)." *Journal of Geophysical Research,* vol. 94, pp. 18265-18271.

Cofer, W.R., III, J.S. Levine, E.L. Winstead, and B.J. Stocks. 1991. "New Estimates of Nitrous Oxide Emissions from Biomass Burning." *Nature,* vol. 349, pp. 689-691.

Crutzen, P.J. 1991. "Methane's Sinks and Sources." *Nature,* vol. 350, pp. 380-381.

Dickinson, R.E., and B. Hanson. 1984. "Vegetation-Albedo Feedbacks." *Climate Processes and Climate Sensitivity,* Geophys. Monograph 29, Maurice Ewing, ed., vol. 5, pp. 180-186.

Dickinson, R.E., G.A. Meehl, and W.M. Washington. 1987a. "Ice-Albedo Feedback in a CO_2-Doubling Simulation." *Climatic Change,* vol. 10, pp. 241-248.

Dickinson, R.E., P.J. Sellers, and D.S. Kimes. 1987b. "Integration Errors in a Three-Dimensional Model for Canopy Albedos." *Agricultural and Forest Meteorology,* vol. 40, pp. 177-190.

Drake, B.G. 1989. *Effects of Elevated Carbon Dioxide on Chesapeake Bay Wetlands. IV. Ecosystem and Whole Plant Responses April-November, 1988.* Report ER-76 (Washington, DC: U.S. Department of Energy).

Fajer, E.D., M.D. Bowers, and F.A. Bazzaz. 1989. "The Effects of Enriched Carbon Dioxide Atmospheres on Plant-Insect Herbivore Interactions." *Science,* vol. 243, pp. 1198-1200.

Fritts, H.C. 1966. "Growth Rings of Trees: Their Correlation with Climate." *Science,* vol. 154, pp. 973-979.

Gornitz, V., S. Lebedeff, and J. Hansen. 1982. "Global Sea Level Trends in the Past Century." *Science,* vol. 215, pp. 1611-1614.

Hansen, J., V. Gornitz, S. Lebedeff, and E. Moore. 1983. "Global Mean Sea Level: Indicator of Climate Change?" *Science,* vol. 219, pp. 996-997.

Hansen, J., G. Russell, A. Lacis, I. Fung, D. Rind, and P. Stone. 1985. "Climate Response Times: Dependence on Climate Sensitivity and Ocean Mixing." *Science,* vol. 229, pp. 857-859.

Hansen, J.E., W. Wang, and A.A. Lacis. 1978. "Mount Agung Eruption Provides Test of a Global Climatic Perturbation." Science, vol. 199, pp. 1065-1068.

Hanson, K., G.A. Maul, and T.R. Karl. 1989. "Are Atmospheric 'Greenhouse' Effects Apparent in the Climatic Record of the Contiguous U.S. (1895-1987)?" *Geophysical Research Letters,* vol. 16, pp. 49-52.

Harper, L.A., D.N. Baker, J.E. Box, Jr., and J.D. Hesketh. 1973a. "Carbon Dioxide and the Photosynthesis of Field Crops: A Metered Carbon Dioxide Release in Cotton Under Field Conditions." *Agronomy Journal,* vol. 65, pp. 7-11.

Harper, L.A., J. Box, Jr., D.N. Baker, and J.D. Hesketh. 1973b. "Carbon Dioxide and the Photosynthesis of Field Crops: A Tracer Examination of Turbulent Transfer Theory." *Agronomy Journal,* vol. 65, pp. 574-578.

Hendrey, G.R., and B. Kimball. 1990. *FACE: Free-Air Carbon Dioxide Enrichment: Application to Field Grown Cotton.* Report BNL 46155 (Upton, NY: Brookhaven National Laboratory).

Hogan, K.B., J.S. Hoffman, and A.M. Thompson. 1991. "Methane on the Greenhouse Agenda." *Nature,* vol. 354, pp. 181-182.

Idso, S.B., B.A. Kimball, and J.R. Mauney. 1987. "Atmospheric Carbon Dioxide Enrichment Effects on Cotton Midday Foliage Temperature: Implications for Plant Water Use and Crop Yield." *Agronomy Journal,* vol. 79, no. 4, pp. 667-672.

Intergovernmental Panel on Climate Change (IPCC). 1990. *The IPCC Scientific Assessment.* Report prepared by Working Group I. Eds. J.T. Houghton, G.J. Jenkins, & J.J. Ephraums (Cambridge University Press, UK, 365 pp).

Karl, T.R., G. Kukla, V.N. Razuvayev, M.J. Changery, R.G. Quayle, R.R. Heim, D.R. Easterling, and C.B. Fu. 1991. "Global Warming: Evidence for Asymmetric Diurnal Temperature-Change." *Geophysical Research Letters,* vol. 18, no. 12, pp. 2253-2256.

Kellogg, W.W. 1991. "Response to Skeptics of Global Warming." *Bulletin of the American Meteorological Society,* vol. 72, pp. 499-511.

Kerr, Richard A. 1983. "El Chichon Climate Effect Estimated." *Science,* vol. 219, p. 157.

_____. 1992a. "Pollutant Haze Cools the Greenhouse." *Science,* vol. 255, pp. 682-683.

_____. 1992b. "Hot Nights in the Greenhouse." *Science,* vol. 255, p. 683.

Kimball, B.A., N.J. Rosenberg, and L.H. Allen, Jr. 1990. "Impact of Carbon Dioxide, Trace Gases, and Climate Change on Global Agriculture." *American Society of Agronomy*, Special Publication No. 53.

Kohlmaier, G.H., H. Brohl, E.O. Sire, M. Plochl, and R. Revelle. 1987. "Modeling Stimulation of Plants and Ecosystem Response to Present Levels of Excess Atmospheric CO_2." *Tellus*, vol. 39B, pp. 155-170.

Lachenbruch, Arthur H., and B. Vaughn Marshall. 1986. "Changing Climate: Geothermal Evidence from Permafrost in the Alaskan Arctic." *Science*, vol. 234, pp. 689-696.

Lamb, P.J. 1986. "An Essay on the Development of Climatic Scenarios for Policy-Oriented Climatic Impact Assessment." Mimeo, International Institute for Applied Systems Analysis (IIASA) workshop.

Lashof, D.A., and D.A. Tirpak, eds. 1990. *Policy Options for Stabilizing Global Climate*. Report to Congress, Main Report PM-221, 21P-2003 (Washington, DC: U.S. Environmental Protection Agency).

Lawrence Livermore National Laboratory (LLNL). 1990. "Systematic Comparison of Global Climate Models." *Energy and Technology Review*, May-June.

Ledley, T.S. 1991. "The Climatic Response to Meridional Sea-Ice Transport." *Journal of Climate*, vol. 4, pp. 147-163.

Lemon, E.R. 1976. "The Land's Response to More Carbon Dioxide." Pp. 97-130 in N.R. Anderson and A. Malahoff, eds., *The Fate of Fossil Fuel CO_2 in the Oceans* (New York: Plenum Press).

Lemon, K.M., L.A. Katz, and N.J. Rosenberg. 1992. "Uncertainties with Respect to Biogenic Emissions of Methane and Nitrous Oxide." Discussion Paper ENR92-03 (Washington, DC: Resources for the Future).

Lincoln, D.E., N. Sionit, and B.R. Strain. 1984. "Growth and Feeding Response to Pseudoplusia Includens (Lepidoptera: Noctuidae) to Host Plants Grown in Controlled Carbon Dioxide Atmospheres." *Environmental Entomology*, vol. 13, pp. 1527-1530.

Lindzen, R.S. 1989. "Greenhouse Warming: Science v. Consensus." Paper (Cambridge, MA: Massachusetts Institute of Technology).

Lindzen, R.S. 1990a. "Some Coolness Concerning Global Warming." *Bulletin of the American Meteorological Society*, vol. 71, pp. 288-299.

_____. 1990b. "Response to A.K. Betts. Greenhouse Warming and the Tropical Water Budget." *Bulletin of the American Meteorological Society*, vol. 71, pp. 1465-1467.

Lough, J.M., and H.C. Fritts. 1987. "An Assessment of the Possible Effects of Volcanic Eruptions of North American Climate Using Tree-Ring Data, 1602 to 1900 A.D." *Climatic Change*, vol. 10, pp. 219-239.

Michaels, P.J., D.E. Sappington, and D.E. Stocksbury. 1988. "Anthropogenic Warming in North Alaska?" *Journal of Climate*, vol. 1, pp. 942-945.

Miller, G.H., and A. de Vernal. 1992. "Will Greenhouse Warming Lead to Northern Hemisphere Ice-Sheet Growth?" *Nature*, no. 355, pp. 244-246.

Mooney, H.A., B.G. Drake, R.J. Luxmore, W.C. Oechel, and L.F. Pitelka. 1991. "Predicting Ecosystem Responses to Elevated CO_2 Concentrations." *Bioscience*, vol. 41, no. 2, pp. 96-104.

Mosier, A.R., and G.L. Hutchinson. 1981. "Nitrous Oxide Emissions from Cropped Fields." *Journal of Environmental Quality*, vol.10, pp. 169-173.

National Academy of Sciences. 1975. *Understanding Climatic Change: A Program for Action*. Report of the Panel on Climatic Variation, U.S. GARP Committee (Washington, DC: National Research Council, National Academy Press).

_____. 1991. *Policy Implications of Greenhouse Warming* (Washington, DC: National Academy Press).

National Research Council. 1989. *Ozone Depletion, Greenhouse Gases, and Climate Change.* Proceedings of the Joint Symposium on Ozone Depletion, Greenhouse Gases, and Climate Change held at the National Academy of Sciences, March 23, 1988 (Washington, DC: National Academy Press).

Neftel, A., E. Moor, H. Oeschger, and B. Stauffer. 1985. "Evidence from Polar Ice Cores for the Increase in Atmospheric CO_2 in the Past Two Centuries." *Nature,* vol. 315, pp. 45-47.

Nierenberg, W., R. Jastrow, and F. Sertz. 1989. "Global Warming Science Flawed?" *The Energy Daily,* June 9.

Parry, M.L., T.R. Carter, and N.T. Konijn. 1988. *The Impact of Climatic Variations on Agriculture* (Dordrecht, Netherlands: Kluwer Academic Publishers).

Penner, J.E., R.E. Dickinson, and C.A. O'Neill. 1992. "Effects of Aerosol from Biomass Burning on the Global Radiation Budget." *Science,* vol. 256, pp. 1432-1434.

Ramaswamy, V., M.D. Schwarzkopf, and K.P. Shine. 1992. "Radiative Forcing of Climate from Halocarbon-Induced Global Stratospheric Ozone Loss." *Nature,* vol. 355, pp. 810-812.

Ronen, D., M. Magaritz, and E. Almon. 1988. "Contaminated Aquifers Are a Forgotten Component of the Global N2O Budget." *Nature,* vol. 335, pp. 57-59.

Rosenberg, N.J. 1981. "The Increasing CO_2 Concentration in the Atmosphere and Its Implication on Agricultural Productivity. I. Effects on Photosynthesis, Transpiration and Water Use Efficiency." *Climatic Change,* vol. 3, pp. 265-279.

_____. 1992. "Adaptation of Agriculture to Climate Change." *Climatic Change,* vol. 31, pp. 385-405.

Rosenberg, N.J., B.L. Blad, and S.B. Verma. 1983. *Microclimate: The Biological Environment* (New York: John Wiley & Sons).

Rosenberg, N.J., B.A. Kimball, Ph. Martin, and C.F. Cooper. 1990. "From Climate and CO_2-Enrichment to Evapotranspiration." Chapter 7 in P.E. Waggoner, ed., *Climate Change and U.S. Water Resources* (New York: John Wiley & Sons).

Schlesinger, M.E., and X. Jiang. 1988. "The Transport of CO2-Induced Warming into the Ocean: An Analysis of Simulations by the OSU Coupled Atmosphere-Ocean General Circulation Model." *Climate Dynamics,* vol, 3, pp. 1-17.

Schneider, S., and R. Londer. 1984. *The Coevolution of Climate and Life* (San Francisco: Sierra Club Books).

Schneider, S.H., and N.J. Rosenberg. 1989. "The Greenhouse Effect: Its Causes, Possible Impacts, and Associated Uncertainties." Chapter 2 in N.J. Rosenberg, W.E. Easterling, III, P.R. Crosson, and J. Darmstadter, eds., *Greenhouse Warming: Abatement and Adaptation* (Washington, DC: Resources for the Future).

Sedjo, R.A., and A.M. Solomon. 1989. "Climate and Forests." Chapter 8 in N.J. Rosenberg, W.E. Easterling III, P.R. Crosson, and J. Darmstadter, eds., *Greenhouse Warming: Abatement and Adaptation* (Washington, DC: Resources for the Future).

Sellers, P.J., and J.L. Dorman. 1987. "Testing the Simple Biosphere Model (SiB) Using Point Micrometeorological and Biophysical Data." *Journal of Climatology and Applied Meteorology,* vol. 26, pp. 622-651.

Shine, K.P., and A. Sinha. 1991. "Sensitivity of the Earth's Climate to Height-Dependent Changes in the Water Vapour Mixing Ratio." *Nature,* vol. 354, pp. 382-384.

Smith, J.B., and D. Tirpak, eds. 1989. *The Potential Effects of Global Climate Change on the United States* (Washington, DC: U.S. Environmental Protection Agency, Office of Policy, Planning, and Evaluation).

Stauffer, B., G. Fischer, A. Neftel, and H. Oeschger. 1985. "Increase of Atmospheric Methane Recorded in Antarctic Ice Core." *Science,* vol. 229, pp. 1386-1388.

Svensson, B.H., J.C. Lantsheer, and H. Rodhe. 1991. "Sources and Sinks of Methane in Sweden." *Ambio,* vol. 20, pp. 155-160.

Tans, P.P., I.Y. Fung, and T. Takahashi. 1990. "Observational Constraints on the Global Atmospheric CO_2 Budget." *Science,* vol. 247, pp. 1431-1438.

Tissue, D.L., and W.C. Oechel. 1987. "Physical Response of Eriphorum Vaginatum to Field Elevated CO_2 and Temperature in the Alaskan Tussock Tundra." *Ecology,* vol. 68, pp. 401-410.

United Nations Environment Programme (UNEP). 1987. *Montreal Protocol on Substances That Deplete the Ozone Layer (*New York: UNEP).

U.S. Department of Energy (DOE). 1990. *Building an Advanced Climate Model: Program Plan for the CHAMMP Climate Modeling Program.* DOE/ER-0479T (Washington, DC: DOE).

Vaghjiani, G.L., and A.R. Ravishankara. 1991. "New Measurement of the Rate Coefficient for the Reaction of OH with Methane." *Nature,* vol. 350, pp. 406-409.

Washington, W.M., and G.A. Meehl. 1984. "Seasonal Cycle Experiment on the Climate Sensitivity Due to a Doubling of CO_2 with an Atmospheric General Circulation Model Coupled to a Simple Mixed-Layer Ocean Model." *Journal of Geophysical Research,* vol. 89, pp. 9475-9503.

_____. 1989. "Climate Sensitivity Due to Increased CO_2: Experiments with a Coupled Atmosphere and Ocean General Circulation Model." *Climate Dynamics* (reprint), vol. 4, pp. 1-38.

Watson, R.T., M.J. Prather, and M.J. Kurylo. 1988. *Present State of Knowledge of the Upper Atmosphere 1988: An Assessment Report.* NASA Reference Publication No. 1208 (Washington, DC: National Aeronautics and Space Administration).

Watson, A.J., C. Robinson, J.E. Robinson, P.J.LeB. Williams, and M.J.R. Fasham. 1991. "Spacial Variability in the Sink for Atmospheric Carbon Dioxide in the North Atlantic." *Nature,* vol. 350, pp. 50-53.

Whalen, S.C., W.S. Reeburgh, and K.S. Kizer. 1991. "Methane Consumption and Emission by Taiga." *Global Biogeochemical Cycles,* vol. 5, no. 3, pp. 261-273.

Wilson, M.F., A. Henderson-Sellers, R.E. Dickinson, and P.J. Kennedy. 1987. "Sensitivity of the Biosphere-Atmosphere Transfer Scheme (BATS) to the Inclusion of Variable Soil Characteristics." *Journal of Climatology* (reprint), vol. 26, pp. 341-362.

3

Nonlinearities and Surprises in the Links of Farming to Climate or Weather

Paul E. Waggoner

Somewhat selectively, I examine in this paper (1) nonlinearities relating components of farming to weather, (2) nonlinearities in the crucial factors of drought and runoff of water that affect farming, and (3) nonlinear changes that weather causes in farming as a whole. As Rosenberg makes clear in his paper, climate scenarios remain dogged by much uncertainty and controversy. The route I have chosen sidesteps this uncertainty by concentrating on an assessment of effects rather than on prospects for climatic change.

Although my concentration is on physical nonlinearities, important nonlinearities also appear in the demand for agricultural products. The price elasticity for food products as a whole is less than one, reflecting the nature of food as a necessity (elasticities for individual products that admit substitutes are higher—see, e.g., Samuelson and Nordhaus, 1989). Similarly, there is a nonlinear relationship between food intake and income above subsistence levels. These nonlinearities interact with physical nonlinearities on the supply side to determine the effects of climate change on agriculture and the consequences for human beings.

As described below, the responses of farming to climate are rich in nonlinearities. For example, if moisture falls below the wilting percentage for corn, the nonlinearity at this moisture level renders the sensitivity of corn yields at higher moisture levels irrelevant. Moreover, if nonlinearities are outside the realm of climate examined within a sensitivity assessment but inside the realm of climate that comes to pass, then the nonlinearities can cause surprises. A physical phenomenon that surprises a person with one set of expectations or convictions may not surprise a person with another set. An event is never surprising in itself; it is potentially surprising only in relation to a particular set of convictions about how the world is, and if it is noticed by the holder of that particular set of convictions (see Thompson, Ellis, and Wildavsky, 1990).

Because the links of farming to weather abound with nonlinearities, we can exploit that fact to help inform our understanding of nonlinear phenomena. Nonlinearities create thresholds that allow us to make useful classifications of natural phenomena. To illustrate, consider the categories of climate and vegetation (Thornthwaite, 1939). Classification allows us to visualize zones of vegetation across the planet as climate varies. Natural vegetation types include tundra, broadleaf forest, Mediterranean scrub, prairie, steppe, and tropical rain forest. Although the earth supports acres that blend, say, savannah and rain forest, still the differences are sufficiently clear that categories can be defined. These categories rest on nonlinearities.

In considering the fundamental nonlinearities that underlie farming, note the nonlinearities inherent in Liebig's (1803-1873) Law of the Minimum: (1) When one essential component is limited, any moderate change in the ample supply of another will affect growth little, and (2) growth is proportional to the limited essential component. Although the law would surely apply to many factors, Liebig had fertilizer in mind. Later, in 1905, Blackman wrote of limiting factors in plant physiology in general, and we can think here of limiting weather factors. In either case, limiting factors cause nonlinearity.

FUNDAMENTAL NONLINEARITIES UNDERLYING FARMING

Among the component processes of farming subject to nonlinearities, many are responses of crops to their environment. Concentration on crops is reasonable because they are the primary producers of food, whether it is processed through an animal into meat or not. Concentration on crops is also reasonable because they generally grow exposed to the weather, whereas animals and markets are often sheltered. I shall, nevertheless, show a few nonlinearities that are not responses of crops.

Germination

Crops generally begin a season as seed. In cold soil, the seed lies dormant. Then, when the soil warms above a threshold near 4°C, dormancy ends and germination begins (Feddes, 1971). Above about 4°C, the speed of germination, measured as the reciprocal of the time for half the seed to germinate, rises linearly. The threshold near 4°C, however, shortens the half-time for germination in a nonlinear fashion.

To germinate, seed must be moist as well as warm. There is a nonlinear rise of respiration—that is, oxygen uptake—as seeds are moistened. To illustrate, below about 15% moisture an oat seed is dormant, between 15 and 20% its respiration is slow, from 20 to 30% respiration accelerates rapidly with more moisture, and above 30% moisture scarcely affects respiration (Bakke and Noecker, 1933). Clearly, passing over thresholds of temperature and moisture can switch germination on and off, whereas in other ranges of temperature and moisture, their changes affect germination little.

Enlargement

After a seedling emerges from the soil, its leaves enlarge at a rate determined by the availability of water in the soil. While leaves enlarge rapidly when they are moist, as soils dry, enlargement falls off nonlinearly. That is, the rate of change in enlargement varies inversely with moisture (measured by suction bars, expressed exponentially). In a representation for three plant species (soybean, sunflower, and maize), the percentage enlargement of leaves is related to moisture with an elasticity of -1.8 (Boyer, 1970). In other words, each unit reduction of moisture, in terms of suction bars, is associated with a 1.8% increase in leaf enlargement, both variables being measured logarithmically.

Photosynthesis and Transpiration

Although water expands crop cells, the energy or food for the growth of crops comes from photosynthesis transforming carbon dioxide (CO_2), water, and solar energy into carbohydrate; thus the peculiar double role of CO_2. As a greenhouse gas it may change climate someday, but as the stuff of photosynthesis it forms food today. Naturally, rising CO_2 concentration speeds photosynthesis.

Rising CO_2 concentration speeds photosynthesis rapidly at first and then less and less. Physiologists have established that plants generally fall into the two classes illustrated by what I shall call the maize class and the wheat class. Maize typifies a class of plants physiologists call C_4, and wheat typifies the class they call C_3. In the maize class, photosynthesis gets under way at low CO_2 concentration, rises rapidly to the concentration now in the atmosphere, and then rises little more. In the wheat class, photosynthesis only begins at some 60 parts per million (ppm), but it rises faster than in maize (Akita and Moss, 1973). So the maize class generally yields more today than the wheat class, but enrichment of CO_2 helps the wheat class more than the maize class. The link between CO_2 and photosynthesis is nonlinear—and it also reveals a clue to a future reversal in the productivity race between these photosynthetic classes of crops.[1]

Warming temperature to the neighborhood of 30°C typically speeds photosynthesis, but further warming slows it (Bjorkman, Badger, and Armond, 1980). However, growing the crop in cool or warm temperatures before measurements are taken modifies the nonlinear relation between photosynthesis and temperature differently for various species. Some species adapted to cool temperatures cannot acclimate to warm, some species adapted to warm cannot acclimate to cool, and some species can acclimate. The maize and wheat classes do not generally differ in their response to temperature (Bjorkman, 1975).

Water affects photosynthesis first because (as Rosenberg's paper also notes) the CO_2 for photosynthesis enters leaves through pores called stomata. The pores are open when the bordering cells are turgid, but close when their turgidity lessens. As the soil in pots growing spring wheat dries during three days and the suction of leaves rises, the stomata close, their conductance for CO_2 falls, and their photosynthesis also falls (Frank, Power, and Willis, 1973). The fall of stomatal conductance is at somewhat lower suctions for young, heading plants than for older, filling plants. The photosynthesis of well-watered plants with nearly zero suction and small stomatal resistance is about 40 milligrams of CO_2 per square decimeter per hour. As the stomatal conductance for CO_2 falls, the photosynthesis plummets at suctions near 15 bars.

The stomata that admit CO_2 into foliage and supply photosynthesis also allow water to transpire from the moist interior to the dry atmosphere. As soil dries and suction of soil water around roots increases, transpiration slows nonlinearly (Denmead and Shaw, 1962). The nonlinearity is slight when the water demand is low because, for example, the sky is cloudy. When a bright sun raises demand, however, the nonlinearity is sharp.

Development

Development is the series of changes an organism passes through from an embryonic stage to maturity. A wheat seedling could transpire water freely, assimilate carbon by photosynthesis, and enlarge indefinitely. But it would never feed us if it did not develop from the seedling to the heading or flowering and then on to the grain-filling stages. Weather affects development as well as enlargement and photosynthesis.

Many plants require a cold period to vernalize them, literally making them springlike and causing them to flower. In the case of wheat, a plant growing from unchilled seed does not flower, but chilling seed 80 days at 5°C causes a plant to flower. Like switching on a lamp, chilling throws a switch for flowering, exemplifying nonlinear response.

The speed of passage from emergence to the flowering switched on by vernalization, however, can be varied by the temperature. How temperature regulates development or phenology has been studied for centuries (see, e.g., Reaumur, 1735). In such analyses the influence of temperature is calculated in terms of growing degree-days, calculated by adding the temperature above a threshold, day by day.

Scientists have found that the threshold for growth varies among plants. To illustrate, table 1 provides threshold temperatures for several crops. Simple degree-days for predicting the speed of passage through development are refined by multiplying the sum by the length of the day. Thresholds mean nonlinearity.

Pests

Crops do not grow alone in the fields; they grow with pests. When climate change is foreseen, the specter of worse pests usually arises. The relations between climate and pests are clear, and the northward movement of the overwintering range of an insect pest, for example, can be calculated for scenarios of warming climate (Stinner et al., 1989). The effect of weather on a pest is nonlinear, however. The nonlinearity allows mapping of zones of pest absence and presence. For example, classes of apple scab in Washington State can be

Table 1. Temperature Thresholds for Various Crops

Crop	Threshold,°F (°C)
Spring wheat	37-40 (2.8-4.4)
Peas	40 (4.4)
Oats	43 (6.1)
Potatoes	45 (7.2)
Sweet corn and snap beans	50 (10.0)
Lima beans and tomatoes	50 (10.0)
Field corn	55 (12.8)

Source: Dethier and Vittum (1963).

mapped more in the humid western part and less in the arid central part. Some pests are favored by cool weather and some by warm, some by wet and some by dry. The shift among these species and even kingdoms is the crucial, nonlinear change. Purple nutsedge flourishes in warm wet weather, field bindweed in warm dry weather, quackgrass in cool wet weather, and Canada thistle in cool dry weather.

Frost

Freezing is a perennial hazard for growing fruit, and early frost can prevent the normal maturing of corn and so require artificial drying. Despite some warming of the planet during the past decade or so, freezing has ravaged the northern counties of the Florida citrus belt (Miller and Glantz, 1988). Nevertheless, global warming should generally lengthen growing seasons by making frosts come later in the fall and end earlier in the spring. A predilection to see only harm in change might cause people to overlook the benefit of less frost.

Frost is, of course, a switch that turns off the life of, say, a tomato plant. Because the relation of life and death to temperature above and below freezing is nonlinear, climatologists can pass beyond the abstraction of climate, expressed in somewhat meaningless average temperatures, to present the odds of last frost after dates in the spring and before other dates in the fall (Thom and Shaw, 1958; Havens and McGuire, 1961).

Biomass

Having reviewed the series of nonlinear responses of component processes of crop production, I turn to an integration of the processes in a single relation. One integration is the proportionality between the solar energy absorbed and biomass accumulated by a crop (Monteith and Elston, 1983). To find an integration that reflects weather, however, I turn also to the proportionality between the accumulation of biomass and transpiration. Figure 1 shows increases of biomass and grain yield of wheat with increasing amounts of precipita-

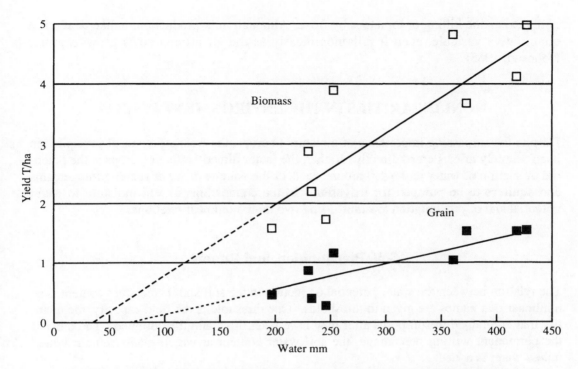

Figure 1. The accumulation of biomass and yield of grain by Kubanka wheat at North Platte, Nebraska, as the amount of water for transpiration increased. (Source: Based on information from deWit, 1958.)

tion and soil moisture. The water is the sum of that stored in the soil plus precipitation. With only 200 to 450 millimeters (mm) of water in the dry climate of North Platte, Nebraska, transpiration essentially equals the supply of water. First the crop must pass a threshold near 30 mm of water to accumulate any biomass. Then it must pass another threshold near 75 mm of water and 0.5 metric tons per hectare (T/ha) biomass to yield any grain.

The nonlinearities in figure 1 are thresholds. Nothing grows below about 30 mm of water, and no grain forms below about 75 mm. The reader should recall the nonlinearities behind the relation of yield to the supply of water for transpiration. The Nebraska wheat grows nicely until the nonlinearity between suction and stomatal conductance chokes both photosynthesis and transpiration. The thresholds and then linear relations between the quantity of water and first biomass, then yield, follow.

Animals

While farm animals are homeothermic, seeking to maintain a constant body temperature, weather does affect them. Heat waves kill poultry in houses, a link of animal to weather as nonlinear as the killing of tomatoes by frost. Although milk production, unlike death, is a quantitative variable, even it falls nonlinearly as the air around cows grows warmer (Johnson, 1965).

NONLINEARITIES IN THE ENVIRONMENT ITSELF

Despite the surpassing importance of moisture to crops, the weather factor of precipitation itself scarcely affects crops directly. Rather, the factor directly affecting crops is the potential or suction of water in the soil around roots or the volume of water running into streams and aquifers to be pumped for irrigation. In the connections of soil moisture to water potential and of precipitation to runoff I find two more nonlinear relations.

Soil, Precipitation, and Runoff

The relation between the water potential or suction in the soil and its moisture content is so nonlinear that names are given to thresholds. One threshold is the field capacity, the quantity that is retained in soil pores after they have been filled and then drained. The other is the permanent wilting percentage, the soil water content at which plants remain wilted unless water is added.

The water Q that runs into streams is the residuum after evaporation E is subtracted from precipitation P. If the amount stored in the soil changes, this simple relation is not quite true, but over a long time, like a year, it is approximately true. Ignoring any change in evaporation with change in precipitation, the elasticity of the runoff with changing precipitation is

$$\text{d} \log(Q)/ \text{d} \log(P) = P/(P - E) = P/Q$$

From the Carolina coast to western Texas, precipitation declines from 1,200 to 400 mm. Accompanying the decline of precipitation is a rise in evaporation from foliage, soil, and open water that can be perceived in the rise of evaporation from open water from 1,200 to 2,400 mm. In fact, Schaake (1990) found the elasticity of runoff for a change in precipitation rose from about 2 on the Carolina coast to more than 4 in western Texas.

Now I turn to the probabilities of different amounts of precipitation. Fortunately, the variances of the frequency distributions of monthly and annual amounts of precipitation are rather closely correlated with average precipitation. From this simplification one can anticipate a change in the frequencies of precipitation below a drought threshold when the average precipitation changes. Elasticities, or relative changes in these drought probabilities with changes in average precipitation, can be calculated as shown in figure 2. The three curves show the percent probability of precipitation being less than thresholds of 6, 25, and 100 mm (1/4 inch, 1 inch, and 4 inches), as a function of the mean precipitation. Low thresholds and high averages make for big elasticities. The elasticity for the probabil-

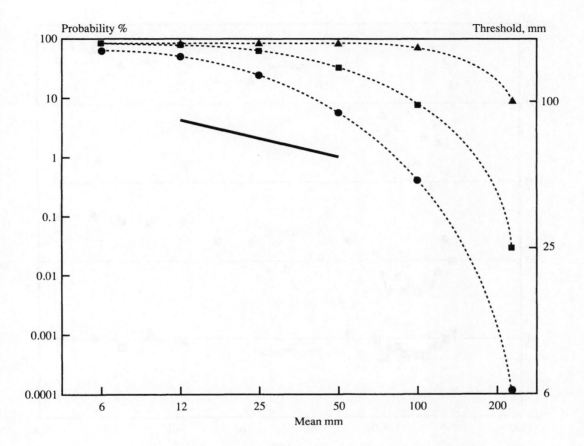

Figure 2. Drought probabilities as functions of mean precipitation for different drought thresholds. The dotted curves show probabilities for thresholds of 100 mm (triangles), 25 mm (squares), and 6 mm (circles) of precipitation, and the slope of the solid line shows an elasticity of 1. (Source: Waggoner, 1989.)

ity of precipitation less than a fourth of the average before the change ranges from 2 for an average precipitation of 25 mm to nearly 5 for an average of 100 mm of precipitation.[2]

NONLINEAR RESPONSES OF FARMING AS A WHOLE TO THE ENVIRONMENT

Margins as Nonlinearity

Just as the margins of types of climate and natural vegetation reflect nonlinearities in the relation of plants to climate, the margins of farming regions or prevalence of a crop show

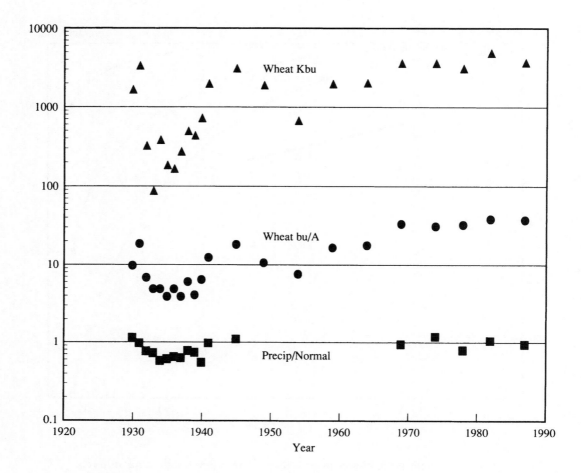

Figure 3. The precipitation and the yields per acre and production of wheat in Haskell County, Kansas, during and after the Dust Bowl period (Kbu = thousands of bushels; bu/A = bushels per acre). (Source: Based on information from Worster, 1979, p. 151, and U.S. Department of Agriculture, various dates.)

nonlinearities in the responses of farming as a whole to the climate. For example, Rosenberg (1982) has taken advantage of the sharp northern margin of hard red winter wheat to show its northward migration and hence its adaptation to colder climates during 60 years.

The response of a system of farming to weather is exemplified by Haskell County, Kansas, during the Dust Bowl years. The county measures 40 kilometers (km) by 40 and it lies about 50 km from the Oklahoma Panhandle and 80 km from Colorado. The precipitation and wheat output of Haskell County for two generations of farmers are shown in figure 3. The vertical scale is logarithmic to show the relative changes in precipitation, yields per acre, and production in the county. Plotted in this fashion, the clear patterns are some

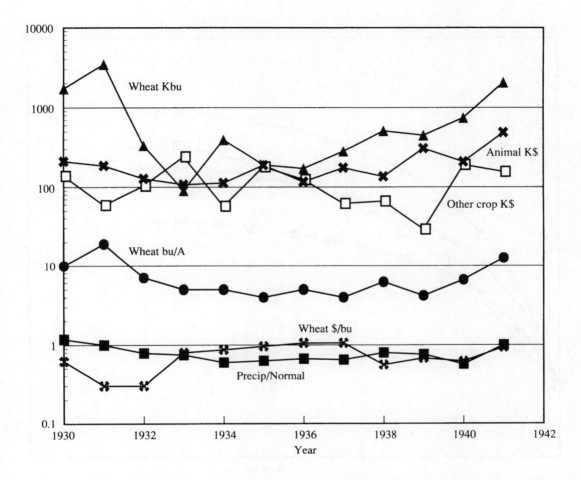

Figure 4. Precipitation and farming in Haskell County, Kansas, during the Dust Bowl period (Kbu = thousands of bushels; K$ = thousands of dollars; bu/A = bushels per acre; $/bu = dollars per bushel.). (Source: Based on information from Worster, 1979.)

variations in precipitation, more variation in wheat yields, and greatest variation in total production of wheat. The Dust Bowl period is enlarged in figure 4. The hierarchy of response noted above is again evident.

Nonlinearity Under Different Climate Scenarios

I turn now from empirical evidence about effects of climate change to a broader consideration of agricultural response under different scenarios. The situations considered are illustrat-

Yield

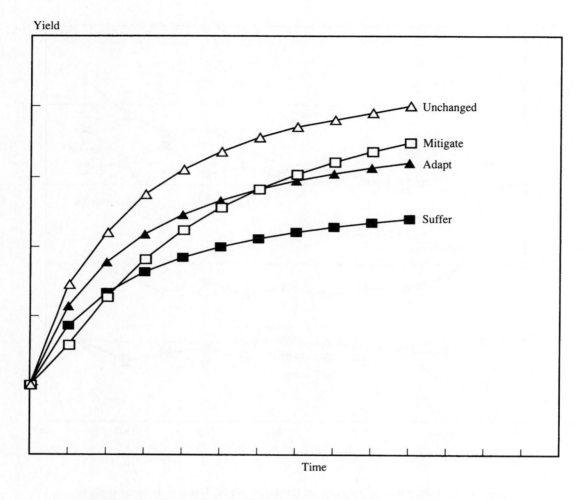

Figure 5. The change in yield of a crop envisioned during the time of a climate change. (Source: Waggoner, 1992. This figure is inspired by a diagram, suggested by Richard Cooper, in National Academy of Sciences [1992, p. 31]).

ed in figure 5. I assume that with no change in climate, advancing technology and other factors would raise yields over time, though at a decreasing rate. Yield thus would follow the path marked "Unchanged" in this case. The harmful effects of climate change are assumed to shift the yield path down to the curve labeled "Suffer." Both adaptation to climate change and mitigation of climate change can eventually reverse some of the losses, as shown.

The nonlinearity in each yield curve reflects the assumption of a slowing increase in yield as some limiting factor in agricultural productivity is confronted. However, I have also allowed for nonlinearities in the incremental effects of adaptation and mitigation over time. There could be a lag in the effectiveness of adaptation. For example, adaptation would involve the introduction of species that have lower yields than the status quo with minor climatic change but are more robust as climate change grows. I have also envisioned that any disadvantage to agriculture of mitigating overall greenhouse gas emissions—say, by lesser use of nitrogen fertilizer—would in time be overcome as mitigation lessened loss of yield by lessening the envisioned harmful change of climate. Because agriculture's contribution to greenhouse warming is relatively small, however, this ultimate lessening of warming and loss of yield would be trivial without other mitigation measures.

SOME USES OF NONLINEARITY

Does the foregoing catalog of nonlinear responses in farming help us anticipate nonlinearities that might otherwise surprise us? I argued previously that clustering of phenomena caused by nonlinearities allows simplification in characterizing effects. In the present context, rather than worrying about the 2,143,150 distinct farms counted by the statisticians in 1990, we can simplify by classification based on distinct regions or types like the Corn Belt or Breadbasket or Northern Hardwoods, all bounded by nonlinearities.

Nonlinearities on a Map

When we map crops, we are actually mapping nonlinearities. The nonlinearities lie on the demarcation lines or margins of cultivation areas. Although investigation would refine the cause of the nonlinearities bounding the regions, it would likely remain temperature latitudinally and precipitation longitudinally in the United States.

The nonlinearities that delineate crop regions also suggest where surprise is likely. Because response is nonlinear, climate can change considerably and affect the yield of spring wheat near Minot, ND, or winter wheat near Wichita rather little. On the other hand, a modest warming could place winter wheat in Montana on the "better" side of a nonlinearity, or a modest drying could place winter wheat in Colorado on the "worse" side. As Parry and Carter (1984) observed, "The focus on marginality in this report derives from the assumption that sensitivity to climatic variability may be more readily observed . . . at the boundaries between different farming systems."

One should not be surprised, of course, if concentrating on change at the narrow margins and neglecting persistence in the broad regions of farming in between exaggerates the expectation of impact from climate change. Unaccountably, the international report of impacts of climate change on farming (Intergovernmental Panel on Climate Change [IPCC], 1990) does not emphasize the advantage of studying the nonlinearities at margins. It does, however, list semiarid—and presumably marginal—regions as among the most vulnerable. It also devotes one of its major findings to the benefit of warming and lengthening of the growing season at the cold margins.

Probabilities

Consideration of drought rather than precipitation, evaporation, and soil capacity, of heat wave rather than hours above some threshold, and of wet spring rather than a complex function of soil, weather, and tractor all simplify thinking. These simplifications rest, of course, on nonlinearities as seen in the relation between the water potential or suction in the soil and moisture content, or in the exemplary nonlinearity of freezing at 0°C.

Once phenomena are classified, rather than measured as continuous quantities, the estimation of probabilities follows readily. When this is done, one ends up dealing with infrequent phenomena on the harmful or beneficial side of a nonlinearity or threshold. Then one encounters a situation where, say, relatively small changes in the average values of rainfall and temperature can have a marked effect on the frequency of extreme levels of available warmth and moisture (IPCC, 1990). For the example of drought, figure 2 explicitly shows how this high elasticity and nonlinearity come to pass.

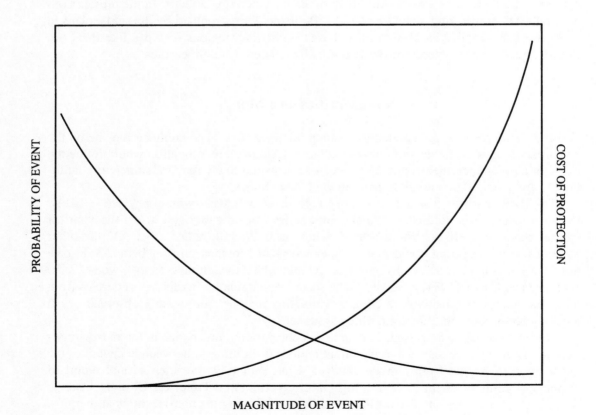

Figure 6. The Planner's Dilemma. (Source: Redrawn, with permission, from Montroll and Badger, 1974.)

Indices Worth Forecasting

Reversing the usual flow of analysis from meteorology and climate scenarios down to biology, economics, and impacts, a panel of biologists and economists recently specified for meteorologists the indices of climate that are relevant to impact assessment and so worth their forecasting (National Academy of Sciences, 1992, p. 503). The analysis in this paper serves a similar purpose for the forecasters of climate. Because of nonlinearity, farming responds rather little to the weather over a fairly wide range. Then, at the point of nonlinearity, a yield or other important variable drops off the edge, as it were.

I therefore suggest that meteorologists need not labor over details but instead concentrate on estimating the probabilities that the climate for a region will fall into one of three large realms: (1) change so little few will care, (2) change a couple or three degrees, requiring adaptation, or (3) change through a significant nonlinearity, making adaptation impractical. After employing nonlinearity to classify climate scenarios into simply inconsequential, practical, or impractical for adaptation, one could then get on with the crucial work of estimating the probabilities of the three classes.

PROFITING FROM NONLINEARITIES DESPITE THE PLANNER'S DILEMMA

A final benefit from the catalog of nonlinearities could be preparing adaptation to them. If we anticipate harm and try to avoid it by adaptation, however, we encounter the final nonlinearity in my catalog, the Planner's Dilemma, as shown in figure 6. Montroll and Badger (1974, p. 136) observed that

> While the probability of an extreme event diminishes rapidly as a function of its magnitude, the destruction it might bring generally rises rapidly with the magnitude as does the cost for protection against it. The planner is confronted with the need to spend more and more money to prepare for an event which becomes less and less likely. [One way of coping is to plan for double the largest observed magnitude of a phenomenon.] This policy is generally an extremely expensive one and usually prepares for an event which might not occur for thousands of years, if ever. Another approach is to assume that an event more than three times the standard deviation from the mean will never occur. The fallacy of this argument is that, if the basic variable is not limited in a natural way, its distribution has a tail which also extends without limit and that, if the sample size is increased, one can expect that the largest value observed will increase . . . . The three standard deviation condition may be far too strong when the sample size is small and too weak when the sample size is too large.

The nonlinearity of the Planner's Dilemma is at the root of the disagreements between those who would stop the world at all costs to avert the chance of frying and those who have other uses for their money than protection against an improbable event. Simply showing the nonlinearities to be faced does not resolve the dilemma. To develop a practical approach I first suggest neglecting inconsequential climate change because it requires no action. I also propose neglecting impossible change because it lies far to the right of figure 6–dangerous and titillating but improbable and profitless. Instead I concentrate on the middle ground, with change of a couple or three degrees, requiring adaptation. As stated by the Council for Agricultural Science and Technology (1992), an important question is, "For a warmer planet with more people, more trade, and more CO_2 in the air, can U.S. farming and forestry prepare within a few decades to sustain more production while emitting less and stashing away more greenhouse gases?" Amidst the scenarios for climate change, answering this question for the realm of a couple or three degrees warming seems an eminently practical course.

How can understanding nonlinearities help lead to answers? Beginning with germination, the clue in its nonlinearity with respect to temperature suggests drawing the margin of a crop zone further north and expecting somewhat fewer fungi damping off the seed. The nonlinearity of freezing provides a similar clue to less drying of green corn in the fall and longer growing seasons and higher yields for, say, potatoes. Adaptations might be matching varieties to new daylengths and managing podzols.

Examining the nonlinearities of enlargement and photosynthesis, one cannot miss the relations to moisture. The preeminent relevant adaptation is coaxing more yield from a water supply that may be lessened by different precipitation and faster evaporation.

The nonlinearities of plant development may require adaptations of vernalization, or they may require varieties tuned to pass through their stages slowly enough to exploit longer growing seasons. One can logically ask if warmer springs might cause early flowers more frequently caught by late frosts than when they bloomed later in cooler springs. Remembering the nonlinearities of pests should lead to devising the management of new pests rather than harking back to old ones. If more frequent heat waves are expected, cheap ways of cooling animals are the priority.

In brief, the probability of a realm of moderate climate change warrants devising adaptations. The need for adaptation underscores the value of searching for clues in the catalog of nonlinear responses of farming to weather.

NOTES

1. An intriguing sidelight in the projected reversal is the issue of whether weeds of the wheat class would become more successful in a maize field as CO_2 rises (Patterson and Flint, 1980).

2. A similar phenomenon changes the probabilities of heat waves relatively more than a change in the average temperature (Mearns, Katz, and Schneider, 1984).

REFERENCES

Akita, S., and D. N. Moss. 1973. "Photosynthetic Responses to CO_2 and Light by Maize and Wheat Leaves Adjusted for Constant Stomatal Apertures." *Crop Science,* vol. 13, pp. 234-237.

Bakke, A. L., and N. L. Noecker. 1933. "The Relation of Moisture to Respiration and Heating in Stored Oats." *Iowa Agricultural Experiment Station Research Bulletin,* 165.

Bjorkman, O. 1975. *Carnegie Institute Washington Yearbook,* vol. 74, p. 748. Cited by Beadle, C. L., S. P. Long, S. K. Imbamba, D. O. Hall, and R. J. Olembo. 1985. "Photosynthesis in Relation to Plant Production in Terrestrial Environments." UNEP Natural Resources and the Environment Series.

Bjorkman, O., M. R. Badger, and P. A. Armond. 1980. "Response and Adaptation of Photosynthesis to High Temperatures." Pp. 233-249 in N. C. Turner and P. J. Kramer, eds., *Adaptations of Plants to Water and High Temperature Stress* (New York: Wiley Interscience).

Boyer, J. S. 1970. "Leaf Enlargement and Metabolic Rates in Corn, Soybean and Sunflower at Various Leaf Water Potentials." *Plant Physiology,* vol. 46, pp. 233-235.

Council for Agricultural Science and Technology. 1992. *Preparing U.S. Agriculture for Global Climate Change.* Task Force Report No. 119 (Ames, IA: Council for Agricultural Science and Technology).

Denmead, O. T., and R. H. Shaw. 1962. "Availability of Soil Water to Plants as Affected by Soil Moisture Content and Meteorological Conditions." *Agronomy Journal,* vol. 54, pp. 385-390.

Dethier, B. E., and M. T. Vittum. 1963. "Growing Degree Days." *New York Agricultural Experiment Station Bulletin,* 801.

deWit, C. T. 1958. "Transpiration and Crop Yields." *Verslagen van Landbouwkundig Onderzoekingen,* vol. 64, no. 6, p. 69.

Feddes, R. A. 1971. *Water, Heat, and Crop Growth* (Wageningen, The Netherlands: Mededelingen van de Lanbouwhogeschool). As shown in J. Wesseling. 1974. *Crop Growth and Wet Soils.* Agronomy No. 17 (Madison, WI: American Society of Agronomy), p. 27.

Frank, A. B., J. F. Power, and W. O. Willis. 1973. "Effect of Temperature and Plant Water Stress on Photosynthesis, Diffusion Resistance, and Leaf Water Potential in Spring Wheat." *Agronomy Journal,* vol. 65, pp. 777-780.

Havens, A. V., and J. K. McGuire. 1961. "Spring and Fall Low-Temperature Probabilities," *New Jersey Agricultural Experiment Station Bulletin,* 801.

Intergovernmental Panel on Climate Change (IPCC). 1990. *Potential Impacts of Climate Change.* Report prepared for IPCC by Working Group II (Geneva: World Meteorological Organization).

Johnson, H. D. 1965. "Response of Animals to Heat." Pp. 109-122 in P. E. Waggoner, ed., *Agricultural Meteorology.* Meteorology Monograph 6 (Boston, MA: American Meteorology Association).

Mearns, L. O., R. W. Katz, and S. H. Schneider. 1984. "Changes in the Probabilities of Extreme High Temperature Events with Changes in Global Mean Temperature." *Journal of Climate and Applied Meteorology,* vol. 23, pp. 1601-1613.

Miller, K. A., and M. H. Glantz. 1988. "Climate and Economic Competitiveness: Florida Freezes and the Global Citrus Processing Industry." *Climatic Change,* vol. 12, pp. 135-164.

Monteith, J. L., and J. Elston. 1983. "Performance and Productivity of Foliage in the Field." Pp. 499-513 in J. E. Dale and F. L. Milthorpe, eds., *The Growth and Functioning of Leaves* (Cambridge, NY: Cambridge University Press).

Montroll, E. W., and W. W. Badger. 1974. *Introduction to Quantitative Aspects of Social Phenomena* (New York: Gordon and Breach Science Publishers), p. 136.

National Academy of Sciences, Committee on Science, Engineering, and Public Policy (COSEPUP) Panel on Policy Implications of Greenhouse Warming, 1992. *Policy Implications of Greenhouse Warming* (Washington, DC: National Academy Press).

Parry, M., and T. Carter. 1984. "Assessing the Impact of Climatic Change in Cold Regions." SR-84-1 (Laxenburg, Austria: International Institute for Applied Systems Analysis).

Patterson, D. T., and E. P. Flint. 1980. "Potential Effects of Global Atmospheric CO_2 Enrichment on the Growth and Competitiveness of C_3 and C_4 Weed and Crop Plants." *Weed Science,* vol. 28, pp. 71-75.

Reaumur, R. A. F. 1735. "Thermometric Observations Made at Paris During the Year 1735, Compared to Those Made Below the Equator on the Isle of Mauritius, at Algiers and on a Few of our American Islands." *Paris Memoirs, Academy of Science,* 1735:545. The reference was found by my former colleague J.-Y. Parlange. The translation, however, is printed in a dittoed copy of a lecture given by C. W. Thornthwaite at the 78th Annual Meeting of the New Jersey Horticultural Society, Atlantic City, December 1952.

Rosenberg, N. J. 1982. "The Increasing CO_2 Concentration in the Atmosphere and Its Implication on Agricultural Productivity. II. Effects Through CO_2-Induced Climatic Change." *Climatic Change,* vol. 4, pp. 239-254.

Samuelson, P. A., and W. D. Nordhaus. 1989. *Economics,* 13th ed. (New York: McGraw-Hill Book Co.), p. 453.

Schaake, J. C. 1990. "From Climate to Flow." Pp. 177-206 in P. E. Waggoner, ed., *Climate Change and U.S. Water Resources* (New York: John Wiley).

Stinner, B. R., R. A. J. Taylor, R. B. Hammond, F. F. Purrington, and D. A. MacCartney. 1989. "Potential Effects of Climate Change on Plant-Pest Infestations." Pp. 8-1 to 8-35 in J. B. Smith and D. A. Tirpak, eds., *The Potential Effects of Global Climate Change on the United States,* Appendix C, Agriculture (Washington, DC: U.S. Environmental Protection Agency).

Thom, H. C. S., and R. H. Shaw. 1958. "Climatological Analysis of Freeze Data for Iowa," *Monthly Weather Review,* vol. 86, pp. 251-257.

Thompson, M., R. Ellis, and A. Wildavsky. 1990. *Cultural Theory* (Boulder, CO: Westview Press).

Thornthwaite, C. W. 1939. "The Climates of North America According to a New Classification." *Geographical Review,* vol. 21, pp. 633-655.

U.S. Department of Agriculture. Various dates. *Agricultural Statistics* (Washington, DC: U.S. Department of Agriculture).

Waggoner, P. E. 1989. "Anticipating the Frequency Distribution of Precipitation If Climate Change Alters Its Mean." *Agricultural and Forest Meteorology,* vol. 47, pp. 321-337.

_____. 1992. "Now Think of Adaptation." *Arizona Journal of International and Comparative Law,* vol. 9, pp. 137-153.

Worster, D. E. 1979. *Dust Bowl* (New York: Oxford University Press).

4

Sensitivity of Unmanaged Ecosystems to Global Change

James S. Clark and Chantal D. Reid

Composition of the atmosphere has changed rather dramatically over the past century. Increased tropospheric concentrations of carbon dioxide, methane, nitrous oxide, and ozone have attended industrialization, forest clearance, biomass burning, and other anthropogenic activities. Atmosphere models suggest that widespread climate change is likely to attend chemical changes in the atmosphere. These effects on climate could become discernible within decades, and some argue that historical data already contain responses to human-caused increases in greenhouse gases. Experiments indicate consequences for plant nutrition, growth, allocation, and water use that, in turn, affect ecosystems in complex ways. Effects on the biota may have already been dramatic, although there is little possibility for observing them.

The many and complex ways in which atmospheric chemistry affects climate and vegetation make difficult any forecasts of their consequences. Variables like temperature, precipitation, light availability, and atmospheric carbon dioxide (CO_2) each affect the biosphere at several spatial and temporal scales. Different scales are associated with different kinds of responses. Climate change affects small-scale phenomena (e.g., microbially mediated nutrient mineralization, gas exchange through leaf stomata) that have consequences for ecosystem function at regional spatial scales. These processes affect humidity and chemical composition of an atmospheric environment shared by other organisms. How climate and atmospheric chemistry effects cascade through ecosystems and how ecosystem responses feed back on the physical environment will determine where and when highly nonlinear responses will occur.

Our goal is to identify some of the key processes that render several of the important North American biomes more sensitive to global environmental changes than others. We begin with a summary of potential responses to changing environment at the ecosystem level. We then provide some interpretations of how these responses may determine sensitivities of the major North American terrestrial biomes to changes in the physical environment. We concentrate on potential effects on vegetation rather than fauna because vegetation types determine habitats for animals, and thus changes in vegetation will determine changes in the rest of the ecosystem. The summary of influences together with their application to major biomes serves as a basis for suggested areas of potentially nonlinear responses to environmental change.

WHAT ARE NONLINEAR RESPONSES AT THE BIOME LEVEL?

Before considering the consequences of global change it is necessary to define the concept of <u>sensitivity</u>, its relationship to system feedbacks, and the way global change can lead to nonlinear responses. <u>Sensitivity</u> is the amount of change in some ecosystem property (e.g., productivity) in response to a change in some factor that affects it (e.g., rainfall). <u>Feedbacks</u> represent interactions with other variables that allow a process to affect itself. <u>Nonlinear</u> responses are those that are disproportionate relative to a change in some driving variable. They are the manifestation of high sensitivity, and they generally involve feedbacks.

Because many processes respond to environmental forces in nonlinear ways, sensitivities vary depending on a number of variables. Consider a simple example. Fire and fuels both depend on climate. In savanna regions and pine forests, absence of fire can lead to changes in fuel structure that change fire probability, severity, and intensity. The altered fire regime then allows for even greater change in vegetation that further influences fire behavior. This is an example of a positive feedback that can potentially translate a rather small change in fire regime into a large one. This scenario was probably realized about A.D. 1550 to 1600 at the prairie-forest border of Minnesota, where a modest increase in moisture availability ultimately translated oak/savanna into mesic "bigwoods" vegetation containing many fire-intolerant hardwoods (Grimm, 1983). The concepts of sensitivity, feedbacks, and nonlinear responses go hand in hand. We seek to identify areas and processes that are highly sensitive to factors likely to change in the coming decades by way of perceived nonlinear dependencies.

WHAT PROCESSES DETERMINE WHETHER AN ECOSYSTEM IS SENSITIVE?

Hydrology

Movement of water through ecosystems is among the most important determinants of plant growth, nutrient cycles, and carbon exchange with the atmosphere. Global change will affect hydrologic cycles in several ways, because atmospheric demand for soil moisture depends on abiotic factors such as temperature, moisture, irradiance, wind speed, atmospheric CO_2 concentration, and plant cover. If growing-season temperatures rise without changes in atmospheric moisture, then demand for soil moisture will increase. Increased atmospheric CO_2 could cause leaf area to increase, resulting in increased interception and water use even though plants may use water more efficiently. It is uncertain whether direct effects of CO_2 enrichment on water use would, in themselves, significantly affect availability, runoff, and/or leaching (e.g., Jarvis, 1989; Eamus and Jarvis, 1989).

Effects of global change on moisture availability and, therefore, ecosystem processes will depend importantly on landscape heterogeneity (topography, current moisture status, and vegetation cover) and temporal variability of new climates. General circulation models (GCMs), used to explore consequences of atmospheric chemistry for the climate system, do not predict spatial variability in moisture status well. One source of uncertainty in the models is the scale at which they simulate climate. GCMs simulate as a single grid point the climate for a large region. Precipitation varies with topography and other factors on much finer scales. Atmospheric demand and precipitation are hard to model, because clouds are patchy at fine spatial scales. Clouds have two different effects on incoming energy: a greenhouse effect and reflection. Accurate climate predictions may depend on achieving the correct balance between these effects at local scales. Given the uncertainty of model predictions, it is difficult to predict how moisture status will change for a given region (Kellogg and Zhao, 1988; Rind, 1988). Without the understanding of regional consequences, it is particularly difficult to speculate on sensitivities at more local scales.

Soils

Soil fertility has profound consequences for agricultural productivity and for the distribution of vegetation. Global climate change could affect soil organic matter, soil erosion, and weathering rates. Erosion rates will accelerate most dramatically in poorly vegetated ecosystems, such as arid lands where land use is intense, or in areas characterized by soils that are shallow (e.g., many mountainous areas) or have low infiltration rates (e.g., compacted soils). Weathering rates could increase with changes in precipitation and in vegetation that alter litter chemistry, temperature, and moisture, potentially being most pronounced in relatively young soils of glaciated regions (e.g., Walker, 1991). Changes in soil temperature could have longer-term consequences for soil chemistry and fertility. Climate-induced changes in soil properties would develop gradually. Without better estimates of how water balance may change, it is difficult to identify areas of sensitive soils.

Nutrient Cycling

Availability of mineral forms of nitrogen (N) and phosphorus (P) limits plant growth on many parts of the globe. A large portion of the annual supply of these nutrients in many ecosystems comes from decomposition of organic matter. Climate change could affect the cycling of mineral nutrients through natural ecosystems through its influence on the rate of nutrient release from decaying organic matter.

Decomposition of organic matter is a key step in many nutrient cycles, and it depends importantly on temperature and moisture availability (Vitousek, 1982). Increasing temperatures can accelerate mineralization (release of mineral nutrients from decaying organic matter), potentially increasing availability of mineral N and P in the short term (e.g., Van Cleve et al., 1990). Large soil organic-matter reserves at high latitudes reflect the fact that decomposition rates decline with decreasing temperature more severely than do rates of litter production. Rising temperatures will therefore reduce soil organic matter in cold soils. By itself, higher temperature would result in a net release of carbon (C) to the atmosphere and of nutrients.

Moisture availability also affects soil nutrient cycling. Soil microbial biomass (Zak et al., 1991), respiration rates (Insam, 1990), mineralization rates (Clarholm et al., 1981), and decomposition (Meentenmeyer, 1978) are correlated with water balance, suggesting they would be affected by climate change. Nitrogen can be lost from soils through denitrification, a process that will change in importance in areas of poor drainage (Boul et al., 1990). Sensitivity of decomposition rates to moisture is probably highest where soil moisture is low (Parton et al., 1987). Pastor and Post (1988) point out that soils with low moisture produce vegetation with poor-quality litter that results in vegetation with still lower quality. This feedback between plants and soil nutrient availability could provide for unexpected effects of climate change on net primary productivity (NPP). Walker (1991) argues that poor soils where NPP is boosted by increased moisture availability could experience elevated C:N ratios in soil organic matter, as litter quantity increases more in response to increased moisture than does litter quality.

Atmospheric CO_2 might have important effects on nutrient cycling through its effects on litter quality. Plant tissues developed under CO_2 enrichment have higher C:N ratios (Luxmoore, 1981; Norby et al., 1986a, b), increased structural C (Oechel and Strain, 1985), and lower lignin (Norby et al., 1986b). The composite effects of altered litter chemistry on mineralization rates are unclear because decomposition rate decreases with increased C:N ratio, whereas lower lignin content favors more rapid decomposition.

Natural Disturbance

Natural and human-caused disturbances have large impacts on managed and unmanaged ecosystems. Given forecasts that convection storms are more likely in the future, lightning ignition of fires and blowdowns of trees could increase. Increased drought frequency or reduced relative humidity with elevated CO_2 (Beer et al., 1988) could increase fire hazard in some ecosystems. Although disturbances can increase rates of successional change (Abrams and Scott, 1989) and accelerate climatically forced compositional changes (Overpeck et al., 1990), it is not clear whether such disturbances will actually become more frequent. Dry lightning storms might have to increase substantially in order to realize a significant increase in fire frequency, because most fires in many areas are started by humans (e.g., Haines et al., 1975). The increase in droughts or unusual weather patterns associated with catastrophic fires might have a more important influence on fire in many regions. Changes in fire regime could have important implications for species composition and ecosystem properties of semiarid regions supporting savanna, pine, and woodland environments (Clark, 1993).

More profound changes in fire regime will attend invasions of exotic grasses in subtropical areas (D'Antonio and Vitousek, 1992). Once grasses invade woodland understories, rapid recovery of a continuous fuel bed following fire is possible. Frequent surface fires become the norm in areas where fire was previously uncommon. Fires perpetuate grass dominance and exclude native species, often changing the entire structure and composition of large areas. Most sensitive are xeric areas where grasses can gain an initial foothold and fuels remain dry long enough to permit fire.

Damage to standing trees during storms (<u>windthrow</u>) can be substantial in natural forests. Because tree size and canopy architecture are important in determining windthrow, it is difficult to speculate on the potential for increases in blowdowns on the basis of climate forecasts. Large trees emergent above the surrounding canopy and trees at forest edges are most susceptible to windthrow (Foster, 1988). Actual losses to windthrow if the incidence of gusty winds increases will be largely controlled by forest structure, which depends in turn on past and present land use.

Insect Herbivores

Global change could affect the role of herbivory in several ways. First, CO_2 and water balance could change the availability and quality of host plants, which in turn will affect insect distributions. Reduced nutritional quality of host plants, which contain higher tissue C:N when grown under elevated CO_2, could result in higher feeding rates (Lincoln et al., 1986). If plants use the extra absorbed C to make greater quantities of defensive compounds, insects could respond by altering their diets. Altered feeding patterns could affect insect development and survival (Lincoln and Couvet, 1989).

Second, global change affects herbivores in more direct ways. Temperature is among the most important influences on insect growth rates. Various aspects of insect development rely on temperature cues, thus causing developmental rates to change with climate. Global warming could increase the number of generations per year of some insect species in some areas. Some insects may now only barely attain reproductive stages each year. Inadequate food supply can reduce growth rates, extend times between molts, and eventually affect survivorship. Lower plant nutritional quality that reduces larval growth rates could affect fecundity (Fajer et al., 1991) and, therefore, population dynamics.

Third, global change could influence the abundance of competitors, predators, and parasitoids with which herbivore species interact. These effects involve a network of nonlinear food-web interactions that would be difficult to predict. For example, predators on locusts decline during North African droughts and contribute to plagues. The highly nonlinear interactions that make prediction difficult today also limit abilities to forecast sensitivities to future global changes.

Species Composition and Diversity

Species composition will undoubtedly be affected by global change, but predictions are difficult. Rising temperatures and direct effects of CO_2 enrichment can affect many aspects of "competitive ability" (Strain and Bazzaz, 1983; Williams et al., 1986; Hunt et al., 1991) and permit range expansions into new regions (Sasek and Strain, 1990). A simple assumption of positive growth responses to elevated CO_2 would lead to the prediction of more rapid succession following disturbance (e.g., Botkin et al., 1973), but long-term growth responses to CO2 fertilization are still unknown. Increased temperatures that allow northward expansion of those species limited by minimum winter temperatures or length of the growing season will bring together populations that today rarely interact. Observed correlations between vegetation type and climate would lead one to conclude that particular climate-change scenarios could have dramatic effects on species composition (Emanuel et al., 1985). Water-use efficiencies and drought tolerances will partly determine the assemblages of species that will dominate new moisture regimes. Differential susceptibility to drought could cause species to react differently to changes in the variance in precipitation.

Species will change their geographic distributions in response to large-scale climate change. Rapid movement of tree populations since the last ice age indicates a potential for rapid rearrangement of species range limits (Davis, 1981; Dexter et al., 1987; Birks, 1989; Clark, 1992). Continued expansion of exotic species is a component of global change that could affect many aspects of ecosystem function. Aggressive establishment of introduced grasses has already altered species composition in many areas through effects on fire regimes (D'Antonio and Vitousek, 1992). Forecasting new species combinations in the context of any specific environmental backdrop is difficult, because the effects of climate and CO_2 are channeled through their effects on individual plant growth and allocation schedules (i.e., life history).

Productivity

Many of the responses to global change discussed thus far have implications for plant production. Increased length of the growing season is likely to result in higher productivity in regions where it is currently limiting. Changes in canopy architecture and leaf duration could affect productivity as a consequence of altered photosynthetic capacity and prolonged activity into cooler seasons. Increased moisture availability could increase NPP anywhere moisture is limiting.

CO_2 fertilization effects are difficult to predict because many plants are not carbon-limited. The way in which productivity responds to rising CO_2 depends on a number of factors. Strong interactions between temperature and CO_2 are likely at the canopy level (Long, 1991). Short-term increases in whole-stand production rates are likely as assimilation rates of individual trees increase (e.g., Solomon, 1986). The degree of acclimation to CO_2 enrichment in mature trees, however, is not well understood (Jarvis, 1989). The response is expected to depend on ability of plants to use increase absorbed C, which depends in turn on moisture availability, temperature, nutrients, other environmental stresses, stand age, and genetic potential (Oechel and Strain, 1985). The higher assimilation rates possible under CO_2 enrichment raise the question of whether more shaded leaves in the lower canopy might experience positive C balances, thereby increasing sustainable leaf area (Jarvis, 1989). This positive feedback might further elevate stand assimilation rates. Maintenance respiration is roughly proportional to plant mass, and it increases linearly to exponentially with temperature. The tendency for elevated temperature to increase maintenance respiration can partially offset any increases in assimilation rate that occur with elevated CO_2 and/or temperature. Maintenance respiration becomes increasingly important with standing crop, suggesting it as a potential sink in forests, particularly boreal forests (Waring, 1991), where temperatures might rise together with CO_2 concentrations.

HOW SENSITIVE ARE THE MAJOR NORTH AMERICAN BIOMES?

Arctic

Productivity and species composition in tundra depend in part on (i) temperature and precipitation gradients, (ii) nutrients, (iii) topography, (iv) length of the growing season, and (v) light. Temperature and moisture affect productivity through their control on thaw depth and snowpack, the distribution of organic soils, and decomposition. These climate effects are mediated by topography, which controls patterns of drainage and heat. Wet tundra is characterized by a moist active layer over permafrost. The wet layer is caused by poor drainage above the permafrost and low evapotranspiration (< 250 millimeters per year [mm yr-1], Stephenson, 1990). Nutrient availability is highly heterogeneous; concentrations of nutrients can vary by an order of magnitude between adjacent microsites (Chapin and Shaver, 1985a). This heterogeneity probably reflects high sensitivity to variability in local moisture conditions. As a result, plant growth is closely associated with gradients in soil moisture and nutrient availability. Growth is limited in part by the degree of leaf area development because of the short growing season (Chapin and Shaver, 1985b). Even though irradiance is low at high latitudes, long daylight hours during the growing season result in low respiratory C losses (Tieszen, 1975).

Climate Sensitivity

Temperature exerts strong direct and indirect controls over arctic ecosystems, and climate models generally agree that the largest temperature increases will occur at high latitudes. With increasing temperature, melting permafrost will affect below-ground processes. Increased depth of the active layer could alter evapotranspiration rates, drainage patterns, rooting depth, and nutrient cycling. In permafrost, organic matter is protected from decomposition. Mineralization and respiration losses of CO_2 derive from the active layer, so increasing its depth is expected to result in increased nutrient availability and a net return of C to the atmosphere (Billings et al., 1982, 1983). The sensitivity of decomposition processes to water table, together with spatial variability in nutrient pools, suggests that nutrient availabilities will increase with increased temperatures. Enhanced nutrients resulting from temperature increase could increase productivity.

Changes in length of the growing season and in irradiance will affect productivity and phenology. Regardless of their adaptations to low temperature, arctic plants must still cope with a short growing season. Temperature increases that hasten snowmelt or delay first snowfall would extend the growing season and potentially enhance productivity. Irradiance is also important for C gain at high latitudes, where a short growing season limits the time plants can engage in photosynthesis, and solar radiation is low even in daylight. GCM-predicted increases in cloudiness could reduce photosynthesis (Grulke et al., 1990). Together, changes in temperature, growing season length, and light availability could alter competitive relationships among plant species and lead to new species combinations.

Dramatic vegetation changes since the ice ages support the notion that climate change would affect arctic ecosystems. The geologic record and GCM modeling experiments suggest that different combinations of climate variables from those of today are possible with changes in global circulation (Kutzbach and Wright, 1985). Xeric "steppe tundra" assemblages of the full glacial, which contained *Artemisia,* grasses, and sedges, occupied Beringia through the northern Yukon and lined the southern margin of the North American ice sheet. Steppe tundra at the southern ice margin might reflect cold, dry easterlies produced by high pressure over the ice (COHMAP, 1988) that also maintained a steep temperature gradient. Composition and productivity of xeric tundra is poorly understood, as they do not appear to match modern arctic assemblages (Lamb and Edwards, 1988). In Fennoscandia and Siberia and northwest of the North American ice sheet, trees and thermophilous taxa grew well north of their modern limits, presumably reflecting higher summer temperatures (Ritchie et al., 1983; Lamb and Edwards, 1988). Thus, different combinations of temperature and precipitation seasonality could produce future assemblages unlike those observed in the current arctic. Given that interglacials like our present Holocene are short-lived (10^4 yr) relative to glacial episodes (10^5 yr), modern wet tundra represents the expansion of a vegetation type uncharacteristic of the last 2 million years of North American history. The extensive wet tundra of the present day is a recent phenomenon.

Arctic ecosystems have adjusted to naturally occurring climate changes of the past. Modern climate-vegetation relationships, together with GCM predictions of increased temperatures at high latitudes, suggest that arctic environments may be less extensive again in the future (e.g., Emanuel et al., 1985), and they may differ from modern tundra in terms of combinations of species and ecosystem processes such as productivity and nutrient cycling.

CO_2 Sensitivity

Existing evidence suggests that direct effects of rising CO_2 levels on arctic ecosystems may be modest. Experiments in upland tussock tundra show little long-term effect of elevated CO_2 on photosynthesis, growth, evapotranspiration, or water use. Photosynthetic rates of the dominant sedge *Eriophorum vaginatum* (Tissue and Oechel, 1987) and whole-system C gain (Grulke et al., 1990) adjusted to artificially elevated CO2 concentrations within weeks. CO_2 enrichment had no significant effect on respiration rates in upland tussock tundra, even when a concentration of 680 parts per million (ppm) was combined with a 4°C temperature increase (Oechel and Riechers, 1986). Although plants produced more tillers under CO_2 enrichment, new tillers were small enough that growth and standing crop were not significantly affected after one year. By the third year, sedge density increased relative to shrubs (Oechel et al., 1991), suggesting that changes in species composition and community structure might eventually become important. Lack of a long-term increase in photosynthetic rates and growth might reflect nutrient limitation or other environmental factors (reviewed by Mooney et al., 1991; Bazzaz, 1990). Evapotranspiration did not respond to elevated CO_2, so water use was also unaffected, probably because water is non-limiting, and stomatal conductance of these arctic species is relatively insensitive to CO_2 concentration (Tissue and Oechel, 1987).

Belowground respiration rates in upland tussock tundra may not be particularly sensitive to CO_2 increases. Elevated CO_2 had little effect on soil C storage in microcosm experiments that consisted of soil cores from upland tussock tundra incubated at doubled current atmospheric CO_2 concentrations (Billings et al., 1983). Although upland tundra may now be a net source of C to atmosphere, stimulation of productivity by CO_2 enrichment is unlikely to have any sustained effect on C balance, particularly if cloudiness increases (Grulke, et al. 1990) as predicted by some GCMs. Tundra is likely to remain a net source of CO_2 to the atmosphere. Direct measurements of CO2 release in dry upland as well as wet lowland tundra support this hypothesis (Oechel et al., 1991).

Water-table effects appear to be more important for C storage than are ambient CO_2 levels. Increases in C capture at CO_2 concentrations of 800 ppm were minor compared with C losses caused by peat decomposition with a 10-centimeter (cm) lowering of the water table (Billings et al., 1983). Increased decomposition would also release nutrients, but Billings et al. concluded that any resultant increases in productivity would not offset the C losses due to respiration.

Boreal Forest

Boreal forest in North America lies between the summer and winter positions of the arctic front (Bryson, 1966). The northern tree line coincides approximately with the low-latitude limit of continuous permafrost. Winters are long and cold and have short days with continuous snow cover. Most precipitation falls during cool summers, when days are long and solar radiation low. Total annual precipitation averages 250 to 600 mm yr^{-1}, but annual deficits tend to be low (Stephenson, 1990). Water balance may influence competitive relationships and fire regimes. Within the boreal zone, influences of topography on heat balance, and thus soil temperatures, exert strong control over permafrost distribution, drainage, nutrient cycles, and vegetation patterns (Van Cleve et al., 1990). Conifers and early successional hardwoods dominate the tree layer of closed canopy forests or of open woodlands with lichen or feathermoss ground cover. Stands are characterized by low nutrient-cycling rates because of low soil temperatures under dense canopies and production of low-quality litter that decomposes slowly (Van Cleve et al., 1990). Fires are common, returning at intervals of several decades to centuries (Heinselman, 1981).

Climate Sensitivity

Temperature. Rising temperatures might have important effects on boreal ecosystems directly through effects on plant physiology, microbial activity, and hydrologic processes and indirectly through lengthening of the growing season. For example, photosynthesis and respiration are both sensitive to temperature. Warmer temperatures would enhance productivity by extending the length of the growing season and by increasing photosynthetic rates. Growth rates (Bonan and Shugart, 1989) and growth form (Payette et al., 1985) of black spruce are especially sensitive to temperature near the northern limit of forest.

Large soil organic-matter pools make boreal forests particularly sensitive to increased decomposition that would attend rising soil temperatures. With increased air temperatures, increased depth of the active layer in permafrost areas, accelerated decomposition, elevated N and P availability, and increased C losses from soil organic matter are expected (Van Cleve et al., 1990). Soil-heating experiments in a black-spruce/feathermoss community near Fairbanks, Alaska, showed evidence of permafrost thaw followed by increases in concentrations of mineral nutrients in soil solution and leaf N concentrations in black spruce. Such increased nutrient availability might enhance growth responses to elevated CO_2 and/or temperature. However, as for the arctic, the increased C sink in aboveground production may not offset C losses due to decomposition.

Several factors are expected to interact with temperatures to produce responses that are difficult to predict. Topography will control the effects of rising temperatures on soil processes, because of its effects on solar radiation. The shading effect of closed forest canopies means that vegetation cover will likewise moderate temperature effects. Lichen and moss layers affect soil temperatures, water relations, nutrient cycling, and tree recruitment. Each of these factors complicate the transient responses of boreal forests to changes in the physical environment.

Soil profile development in boreal forests could be altered by rising temperatures; young, high latitude, glacial soils are most susceptible to climate change (Jenny, 1980; Walker, 1991). Moisture availability, litter chemistry, decomposition, and percolating organic acids would be affected by temperature in ways that might result in enhanced profile development. Walker (1991) predicts that temperature effects on soil will be most important in the northern high latitudes, where soil organic matter is highest.

Temperature change will produce other dramatic changes in boreal ecosystems as a consequence of species-specific sensitivities to minimum temperatures. If minimum winter temperatures increase, the southern limit of boreal forest is susceptible to invasion by broad-leaved deciduous trees (Emanuel et al., 1985), because broad-leaved deciduous species could then survive the milder winters (Sakai and Weiser, 1973). This transition from conifers to hardwoods could have profound effects on C dynamics, nutrient cycling, and hydrology, due to fundamental differences between these ecosystems in litter chemistry, phenology, and water use. Replacement of boreal forest will result in large C releases to the atmosphere, because boreal forests store more C than do deciduous forests (Schlesinger, 1993).

Past temperature fluctuations demonstrate that boreal forest is also sensitive to climate change near the northern limit, because success of young seedlings depends on temperature. Along the east coast of Hudson Bay, the latitude of tree line has not moved significantly during recent centuries, but density of individuals below tree line has increased (Payette and Filion, 1985), and tree line has increased in altitude due to a warming trend of the last century. In northern Sweden, climatic amelioration that occurred during the 1870s and during this century has favored Scots pine, thereby prohibiting development of "climax" Norway spruce in this region (Steijlen and Zackrisson, 1987). Thus, species composition below tree line would change, and the elevation and location of tree line could change with rising temperatures.

Precipitation. Spatial and temporal patterns in species abundances suggest that precipitation changes could affect species composition in boreal forests. Local-scale spatial variability in vegetation with topography reflects the importance of moisture-holding capacity, permafrost distribution, drainage, and depth to the water table. This local complexity will control landscape responses to changing precipitation.

The eastward increase in fir abundance across Canada likely reflects a broad gradient in moisture availability and its effect on fire occurrence. Past assemblages containing spruce and fir further suggest moisture as one determinant of relative abundances. The full-glacial spruce woodland of eastern and central North America contained fir in low abundance, perhaps due to low moisture availability (Watts and Hansen, 1988). A fir increase in eastern North America from 15,000 to 12,000 years ago appears to have followed a rise in precipitation (Prentice et al., 1991). The way species composition will respond to precipitation change will be complicated by differential susceptibility of spruce and fir to fire and/or soils. For example, rather sudden declines in fir abundance in eastern Labrador 6,000 years ago might reflect increased importance of fire or accumulation of humus, either of which might reflect change in moisture availability (Lamb, 1980; Engstrom and Hansen, 1985). Precipitation changes may therefore have direct and indirect effects on species composition within modern boreal forests.

Cloudiness. Long days with low sun angles, short growing seasons, and dominance by relatively unstable Pacific air during the growing season make C assimilation in boreal forests potentially sensitive to changes in cloudiness. The closed canopies of boreal forests make them sensitive to changes in light availability, because evergreen leaf display and canopy architecture is such that photosynthetic rates of most leaves are below the potential. Although individual leaves near the top of the canopy show little response to irradiance once above their saturation point, the larger fraction of the canopy that is light-limited contributes more production with increasing light (Jarvis and Sandford, 1986).

Fire and Insects. Each of the climate variables mentioned above could affect boreal forests indirectly through its influence on fire. At a subcontinental scale, higher fire frequency and tighter fire-weather correlations in drier parts of Canada indicate a high sensitivity of fire to climate change (e.g., Flannagan and Harrington, 1988). More moist forests of eastern Canada support fire regimes that appear less sensitive to climate change.

At local scales, fire is responsible for much of the spatial and temporal pattern in species composition (Johnson, 1979; Foster, 1982) and ecosystem properties (Van Cleve et al., 1991) of boreal ecosystems. Crown fires in spruce and fir (Van Wagner, 1978) and surface fires in pine-dominated stands (e.g., Zackrisson, 1977; Englemark, 1987) affect mortality, regeneration, species composition, active-layer depth, soil organic matter, and nutrient pools. By opening the canopy and blackening soils, soil temperatures are increased, allowing for increased active-layer depth, decomposition, nutrient availability, and C release to the atmosphere (Van Cleve et al., 1991).

The complex interaction of preburn vegetation characteristics, topographic effects, weather, and differing recruitment strategies of postfire vegetation frustrates efforts to forecast effects of climate change on fire regime at local scales. A crude statement can be made that increased temperature and/or lower precipitation could result in higher fire frequency and thus alter species composition and productivity. The changes in fire frequency that appear to have attended past climate fluctuations, together with latitudinal (Payette et al., 1989) and longitudinal (Flannagan and Harrington, 1988) trends in fire importance across the boreal zone support the interpretation that fire regimes are sensitive to climate variability. The large number of factors contributing to the fire-climate-vegetation interaction make detailed predictions difficult.

Still more difficult to predict are the responses to climate change of insect pests, such as the spruce budworm in eastern Canada and Maine, and their effects on the vegetation. Changing seasonality of temperature and precipitation could have consequences for development rates, survivorship, and fecundity of insect populations and their predators. For spruce budworm, cool summers do not permit completion of the life cycle. Increasing temperature may accelerate insect development rates and affect defoliation levels. Large areas of balsam fir can be rapidly lost to spruce budworm, greatly increasing susceptibility to catastrophic crown fire (Ludwig et al., 1978). Small changes in climate could have large effects on insect pests and, therefore, forests because of the highly nonlinear dynamics of some of these pest species. The episodic nature of insect outbreaks makes predictions difficult.

CO_2 Sensitivity

In the absence of climate change, boreal forests are expected to show at least transient growth increases with rising CO_2 (Higginbotham, 1983; Brown and Higginbotham, 1986). This enhancement may be short-lived, however, because other factors may become limiting. In fact, little is known concerning direct effects of CO_2 enrichment in boreal areas. Some correlative evidence suggests direct responses to CO_2. The increased amplitude of the seasonal atmospheric CO_2 oscillation that correlates positively with latitude is one indication that increased photosynthesis, and therefore growth of boreal forest trees, may already have occurred (D'Arrigo et al., 1987). The oscillation might also reflect a decline in forest respiration during winter dormancy. Recent higher temperatures might have played a role, assuming respiration did not increase in step with assimilation (Jarvis, 1989). Increased mineral N in precipitation might have further contributed to increased assimilation in these N-limited stands.

Low-temperature inhibition of photosynthesis during winter and spring could be alleviated with global warming, because of increased length of the growing season, and thus amplify responses to CO_2 enrichment. In the current atmosphere, respiration increases resulting from raised temperatures would offset any increase in C storage resulting from enhanced photosynthesis at higher temperatures. However, limited data on respiration under CO_2 enrichment indicate reduction in respiration rates with elevated CO_2 (Bunce, 1992; Reid, 1990). Clearly, data are needed from boreal forests to determine relative sensitivities of photosynthesis and respiration to temperature and CO_2.

Because sensitivities to CO_2 are potentially significant, relative competitive abilities may also change as a result of CO_2 enrichment. Changes in species composition are difficult to predict because of complex responses to changing temperature, light availability, fire, and CO_2.

Temperate Forest

Temperate forests occur at mid latitudes with seasonal climates. Broad climate gradients within temperate forests of eastern North America are north-south for temperature and east–west for precipitation. Following this climate pattern, the distribution of temperate forest types in eastern North America lies along a southeast-northwest gradient (Braun, 1950; Greller et al., 1990). Dominant vegetation reflects patterns of moisture availability affected by seasonality of temperature and precipitation, soils, and topography. Elevation plays an important role in the western United States, influencing temperature, precipitation, and water demand. Evaporative demand can be high within temperate forests, but high precipitation maintains relatively low moisture deficits in the eastern United States (Stephenson, 1990). The spectrum of soil parent materials and ages (e.g., glaciated, coastal plain) provides for much local variability in vegetation that is not well predicted on the basis of regional climate. Thus a full range of sensitivities to climate change is expected depending on local conditions.

Climate Sensitivity

Temperature. Temperature change will have different kinds of effects in different regions as a consequence of (i) relaxation of minimum winter temperature constraints on northern range limits, (ii) invasion of temperate forests of lower latitudes by subtropical species better able to exploit warmer climates, and (iii) potentially altered interactions among species near centers of their distributions.

Sensitivities of range limits to temperature are best understood in light of the climatic factors responsible for modern distributions. Minimum winter temperatures may be among the more important controls on species range limits in the poleward direction (Larcher and Bauer, 1981; Woodward, 1987; Arris and Eagleson, 1989). Only conifers and some early-successional hardwoods survive at higher latitudes, both groups tolerating extremely low temperatures. In the absence of late-successional hardwoods, the slow-growing, frost-tolerant, and nutrient-use-efficient conifers, Picea and Abies, dominate at latitudes north of those where most temperate species can survive. Evergreen plants can engage in photosynthesis for a longer period because leaves are present all year (Bradbury and Malcolm, 1978). In spring, evergreen conifers can be putting on new growth as soon as temperatures are favorable, whereas deciduous species are drawing on previous years' C reserves to construct new leaves.

Deciduous broad-leaved species dominate where temperatures do not fall below a supercooling temperature threshold near -40°C. Dormant buds can survive much lower temperatures (Sakai and Weiser, 1973), but stem tissue is damaged (Woodward, 1987; Burke et al., 1991). Frost drought and membrane sensitivity are both potential controls on distributions (Larcher, 1982; Woodward, 1987). Broad-leaved deciduous forest dominates regions having adequate moisture during the growing season and minimum winter temperatures above those at which xylem damage becomes acute. Broad-leaved evergreens do not penetrate latitudes where temperatures fall below -10 to -15°C in Japan and the southeastern United States (Sakai and Weiser, 1973), probably because leaves are sensitive to temperatures below -15°C.

Temperatures also affect recruitment success of some populations at northern frontiers, because young seedlings are much more susceptible to frost damage than are large trees. Low temperatures during periods critical for seed development could be responsible for the northern range limits of *Tilia* in the United Kingdom (Piggott and Huntley, 1981). Decreases in temperature at a decade to century scale might explain regeneration failure of *Pinus ponderosa* in parts of the Intermountain West (Stein, 1988), of gymnosperm forests of New Zealand (Wardle, 1963), and of many species near elevational or latitudinal limits (Kullman, 1983). With global warming, recruitment limitations may be ameliorated in northern latitudes. However, potential increases in temperature fluctuations may increase freeze-thaw cycles in mild winters, and recruitment success may be further reduced by frost damage to young seedlings.

These low-temperature sensitivities mean that the abilities of many species to survive at higher latitudes with elevated minimum winter temperatures are among the more important effects of temperature increases. Moreover, responses of range limits to temperature change might be somewhat predictable, provided that change in extreme temperature fluctuations could be forecast with some confidence. In general, prediction of extremes is difficult, because they need not change in step with changes in average temperatures. Seasonality and interannual variability in temperatures could have important implications for species composition in the future.

While northern population frontiers may be explained by direct responses to minimum temperatures, southern range limits may be controlled to a greater degree by competition (MacArthur, 1972; Woodward, 1987). Above the -40°C low-temperature threshold, broad-leaved deciduous species may have a growth advantage over conifers, in part due to differences in canopy architecture (i.e., shape), leaf characteristics, and leaf display (angle and arrangement). Sprugel (1989) argues that conifer canopy structure and leaf orientation represent an adaptation for maximizing whole-plant C gain during cooler months. Photosynthesis is more sensitive to temperature at high rather than at low irradiance. Broad-leaved plants support a greater proportion of their leaf area near light saturation in the upper crown, and these leaves have assimilation rates that decline rapidly with decreasing temperature. Needle-leaved species possess small leaves at high angles to incident light that are mostly unsaturated. Thus, many leaves are below light saturation, and more light is transmitted to lower canopy layers. These layers are increasingly expensive at higher temperatures, making the shallow canopies of deciduous broad-leaved species advantageous. Although evergreens can maintain higher whole-canopy photosynthesis at lower irradiance because of their architecture, they generally have a lower maximum photosynthetic capacity, particularly at high temperatures (Oechel and Lawrence, 1985). Higher maximum photosynthetic capacity of deciduous broad-leaved trees combined with a more efficient display of photosynthetic tissue at higher temperature, provides a competitive advantage at mid latitudes.

Fossil evidence suggests that changes near the boreal-temperate boundary are likely with climate change. Fossil pollen data in the northeastern United States suggest that spruce has already begun to increase over the last 2,000 years, perhaps due to decreasing temperatures (Gajewski, 1988). If temperatures were to rise, the opposite trend might be expected, with advance of temperate deciduous species near the northern range limit.

Farther south where winter temperatures are moderate, broad-leaved evergreen species may have the advantage over deciduous species, because they exploit a longer growing season. They may also have an advantage over needle-leaved evergreens, because the advantage of needle leaves may be realized only in winter when temperatures are low (Sprugel, 1989). The declining seasonality with decreasing latitude makes the broad-leaved evergreen habit increasingly advantageous. This advantage may explain the mid-Holocene northward expansion of bayhead vegetation in the Southeast with increasing winter temperatures of the time (Watts, 1980; Webb, 1988).

Moisture. Changes in moisture availability will have greatest effects on temperate ecosystems where moisture is in short supply. Moisture deficits arise either because precipitation does not meet demand or because shallow and/or coarse soils provide for low moisture storage. There is an increasing moisture deficit from east to west in the eastern United States that is likely to result in increasing moisture sensitivity. Much of the western United States is characterized by high deficit, with many tree populations bounded between low-elevation moisture limitations and high-elevation temperature limitations (e.g., Agee and Kertis, 1987). Moisture availability increases with elevation, because precipitation tends to be higher, particularly on west-facing slopes, and evaporative demand decreases. Deficits also decline with latitude because of declining atmospheric demand (Stephenson, 1990).

Moisture deficits reflect the seasonality of precipitation and demand. Moisture that becomes available at times when it cannot be used by plants does not serve to alleviate moisture deficit. For example, lack of precipitation during the summer in the western United States probably explains dominance by evergreen species. Changes in seasonality of precipitation can have a large effect on species composition and productivity.

The effects of changing precipitation and atmospheric demand will depend on soils and local topography. Coarse or shallow soils do not store sufficient water to meet demands of moist forest hardwoods, which tend to be more susceptible to drought than pines. Pines often dominate sites that are dry as a result of low precipitation during the growing season or of excessively drained soils (e.g., the southeastern coastal plain, out-wash sands and other coarse deposits in glaciated regions, and shallow soils over bedrock that hold little moisture). Decreases in precipitation or a lower water table would have proportionately greater effects on available moisture on coarse soils, because total moisture available for plant growth is lower.

In sensitive areas, changes in moisture availability will affect population and ecosystem processes. Growth rate and survival probabilities are affected by water availability (Schulze et al., 1987), as reflected in correlations between growth rates and/or species composition and local or regional water balance (Beasley and Klemmendson, 1980; Woodward, 1987; Stephenson, 1990; Cook and Cole, 1991). Seedling establishment of many tree populations can fail in drought years. Differential recruitment susceptibility to prolonged (Wardle, 1963) or periodic (Austin and Williams, 1988) droughts can have important effects on species composition and population structure (Pastor and Post, 1986). Local soil differences may become increasingly important determinants of species composition near ecotones determined by moisture availability (Daubenmire, 1936; Buell and Martin, 1961; Grimm, 1983; Pastor and Post, 1988). Tree-species diversity drops off rapidly at the prairie-forest border (Currie, 1991). Species-specific differences in water-use efficiency and drought tolerance also determine competitive relationships and thus species composition (Bunce et al., 1977).

The effects of low moisture availability on vegetation can be exaggerated by the feedback on N availability. Moisture affects nutrient cycles by influencing N inputs in precipitation (atmospheric N inputs increase with precipitation), mineralization, and immobilization and N output by losses in subsurface flow and runoff and by denitrification. Moisture availability affects nutrient turnover indirectly through its effects on vegetation (which in turn affects litter quantity and quality), mineralization, and immobilization of N contained in litter. Drought-resistant species (e.g., oaks and pines) tend to have poor-quality litter that decays slowly. Availability of N feeds back on the N cycle, because low availability can lead to changes in species composition that include more efficient users of N returning less N in litterfall. The response of N cycling to moisture availability may be nonlinear (Pastor and Post, 1988). These nonlinear effects are likely to be most important on xeric sites where increasing moisture deficits select for drought-tolerant species that have high nutrient-use efficiency and therefore cycle less N. Models of nutrient cycling that contain climate effects on mineralization and immobilization of N also predict that moisture sensitivity is highest where it is in short supply (Parton et al., 1987; Pastor and Post, 1988).

Fossil data support the view that sensitivity to moisture availability is greatest near forest-woodland boundaries. When a more negative water balance prevailed during the mid-Holocene (McAndrews, 1966; Webb et al., 1983) and before A.D. 1600 (Grimm, 1983; Clark, 1990), vegetation changes were most pronounced near the prairie-forest border. The dramatic response was either because climate changes were greatest or because vegetation was most sensitive at these boundaries between vegetation types. The other region of dramatic vegetation change during the mid-Holocene is the southeastern coastal plain, where pine forest rapidly replaced oaks on these sandy soils 6,000 years ago (Watts, 1970). Variation in temperature, precipitation, and seasonality may all have contributed to vegetation change. In contrast, no dramatic vegetation change was observed in temperate forests distant from the prairie-forest border or on fine-textured soils during the mid-Holocene. The most dramatic climate-induced vegetation changes of the Holocene in eastern North America occurred where effective precipitation was low or soils were excessively drained.

Response times of ecosystems to changes in moisture availability near grassland-woodland boundaries can be highly variable and difficult to predict. Few long-term data are available, but one example comes from forest invasion of prairie since the mid-Holocene in central Minnesota. The decrease in soil organic matter in outwash sands following invasion of prairie by *Pinus banksiana* progressed at a rate of 0.0014 ±0.00050% per year in the epipedon (upper 17.5 cm of mineral soil) and 0.00084 ± 0.000274% per year in the rooting zone (upper 114 cm of mineral soil) (Almendinger, 1990). These rates imply half-lives of 500 and 835 years respectively, and they describe declines in percent organic matter from 5.7 to 1.4% and 1.81 to 0.56%, respectively. Thus, thousands of years (perhaps five half-lives) will be required for organic-matter pools to approach equilibria with new moisture regimes.

Fire. Fire regimes are most sensitive to changes in regional climate patterns where effective precipitation is low. Fire becomes increasingly important with decreasing water balance (e.g., from east to west in the eastern United States). This relationship is highly nonlinear, with fire importance rising rapidly near prairie or savanna boundaries. Neither fire nor moisture stress alone appears sufficient to explain prairie-forest boundaries; a combination of low moisture availability and fire is probably required (Grimm, 1983; Medina and Silva, 1990; Menaut et al., 1990). The feedback effect of vegetation on fire regime—through fuel accumulation, structure, and flammability—probably contributes to this nonlinearity. Near the prairie-forest border in Minnesota, fire regimes changed with annual-, decade-, and century-scale changes in climate (Clark, 1989a, b).

Climate change is likely to affect fire occurrence at local scales. In moist forests, wildfire is most important on dry sites resulting from shallow or excessively drained soils. These sites tend to support flammable vegetation, including fire-adapted species. Fire is probably necessary to explain *Pinus banksiana* and *P. rigida* on outwash in the Upper Midwest and Northeast, *P. resinosa* at high elevations in New England (Engstrom and Mann, 1991), *P. rigida* in the southern Appalachians (Harmon, 1982), and *P. palustris* in the Southeast (Garren, 1943). Fires can also occur frequently in more moist situations, partly because of frequent ignitions since European settlement (Haines et al., 1978). However, fire importance is clearly correlated with soil texture and/or depth.

Because of its episodic nature, fire is sensitive to changes in seasonality and interannual climate variability. In moist forests, fires are most likely during drought years (Lutz, 1930; Haines et al., 1978; Heinselman, 1973; Clark, 1989b). Fires following severe droughts may be necessary to explain maintenance of high densities of *Pinus strobus* in stands that would otherwise be dominated by northern hardwoods and *Tsuga* in northwestern Pennsylvania (Lutz, 1930) and in southern New Hampshire (Foster, 1988). Without fire, *P. strobus* might still persist in some areas but at low densities as a gap species (Lutz and McComb, 1935). Droughts have their strongest effects on fire probability when they occur during a fire season (Haines et al., 1978).

The effect of global change on fire regimes in a cultural landscape is difficult to predict, because habitat fragmentation, land management, and incendiary events affect ignition incidence and fire behavior in complex ways. Fire exclusion on sites that experienced frequent wildfire can result in the establishment of less-flammable understories with feedback effects on composition, structure, and successional pathways (Grimm, 1984; Host et al., 1988; Whitney, 1986). In other areas, fire suppression can lead to dead fuel accumulation that results in more intense fires. The size distribution of fires has also been greatly altered by fire-suppression activities that have tended to make fire less frequent but more severe in chaparral (Minnich, 1983) and probably also in temperate forest (e.g., the Yellowstone fires of 1988).

CO_2 Sensitivity

Direct effects of CO_2 enrichment on C assimilation rates and water-use efficiency could influence successional patterns in dry environments (e.g., the Piedmont of the southeastern United States) by alleviating moisture stress on some species more than on others (Tolley and Strain, 1984a, 1985; Sionit et al., 1985). Water-use efficiency seems to be ameliorated for species initially less tolerant of water stress. Early successional species may show important growth responses when in full irradiance (Tolley and Strain, 1984b). Forest species composition is also likely to change, because shade-tolerant species appear to benefit more than shade-intolerant species from elevated CO_2, provided sufficient nutrients are available (Bazzaz et al., 1990). This amelioration response to CO_2 is greatest for the most shade-tolerant species (Bazzaz et al., 1990; Reid, 1990). Schlesinger (1993) points out that the strict application of GCM-predicted temperature estimates suggesting extirpation of *Fagus grandifolia* from the eastern United States (Davis and Zabinski, 1992) is a different scenario from one that acknowledges potential CO_2-fertilization effects on growth of this species (Bazzaz et al., 1990; Reid, 1990).

Nutrient addition may enhance growth of deciduous seedlings under elevated CO_2 (Brown and Higginbotham, 1986; Norby et al., 1986a, b; O'Neill et al., 1987). In nutrient-poor stands, enhanced growth is also observed, but much of that growth is directed below-ground. Greater root-mass development results, with aboveground biomass being correlated with ambient CO_2 (Norby et al., 1986a). In most deciduous species, nutrient concentration is diluted at elevated CO_2, leading to greater tissue C:N, which may reduce litter decomposition rates and thus nutrient availability.

Changes in leaf nutrient concentration affect nutritional quality and thus may enhance insect feeding rates (Fajer et al., 1991). Such increases in feeding rate may cause increased forest defoliation. If so, the effect might be short-lived, because subsequent reductions in growth rate or fecundity might reduce herbivore population densities (Fajer et al., 1991).

Savanna, Grassland, Desert

Arid and semiarid lands are characterized by a negative water balance; that is, precipitation generally falls well short of atmospheric demand for soil moisture during much of the year. Annual deficits in North America are 500 to 1,500 mm for the Central Plains, > 900 mm for cold desert (Great Basin sagebrush), and generally exceed 2,000 mm in warm desert (creosote bush) (Stephenson, 1990). Because precipitation variability tends to increase with decreasing total annual precipitation, drought frequency increases with increasing aridity.

Ecosystem processes of the Central Plains of the United States are closely tied to precipitation gradients (Parton et al., 1987; Sala et al., 1988; Schimel et al., 1990; Burke et al., 1991). Soil moisture storage may be a better indicator of productivity than is total annual precipitation, because storage integrates precipitation inputs over several years, thereby relieving some of the moisture stress that can occur during short-term droughts (Burke et al., 1991). Nutrient limitation may be most important in wet years when moisture is not as limiting (Knapp and Seastedt, 1986). Nutrient availability in arid and semiarid lands is correlated to regional climate, parent material (Scholes, 1990), local site conditions (Schimel, 1988; Schlesinger et al., 1990), and time since fire (Knapp and Seastedt, 1986)(see below).

Climate Sensitivity and Land Use

Changes in moisture availability due to increased evaporative demand and/or decreased precipitation will have important direct and indirect effects on temperate grasslands.

Productivity and/or species compositional responses to changes in precipitation patterns have been observed in vegetation surveys conducted during years of favorable and unfavorable water balance (Albertson and Weaver, 1945; Sala et al., 1988) and from the normalized difference vegetation index of the 1988 drought. Vegetation responses to drought are predicted by calibrated simulation models (Schimel et al., 1990; Burke et al., 1991). Highest sensitivity is likely on soils that store little available moisture.

Belowground processes will play an important role in dryland responses to climate changes. In shortgrass steppe, N mineralization rates depend on precipitation amount and moisture variability associated with wetting-drying cycles (Schimel and Parton, 1986). Simulation models of C and N biogeochemistry across the Central Grasslands suggest sensitivity of soil C pools to temperature, with temperature increases boosting decomposition rates. Increased N availability due to raised decomposition rates may increase productivity. If increased atmospheric demand is not attended by increased precipitation, drought stress might limit any productivity response (Schimel et al., 1990; Burke et al., 1991). Sensitivity of soil C pools to changes in precipitation appear highest on coarse-textured soils (Sala et al., 1988). Relative losses of soil C are expected to be highest on drier sites, but largest absolute losses may occur on moist sites (Burke et al., 1991).

The feedbacks among vegetation, fire, and grazing make semiarid and arid lands particularly susceptible to global change. The losses in soil C forecast for the next 50 years as a consequence of rising temperatures on the Central Plains are small relative to those that attended initial land clearance and cultivation (Burke et al., 1991). Modest changes in climate (Walker, 1991) and the introduction of new species, particularly alien grasses (D'Antonio and Vitousek, 1992), can have important impacts through modification of fire regimes and grazing pressures. Fire and grazing further modify vegetation characteristics to produce dramatic changes in vegetation structure and composition. *Prosopis* expansions that transformed large areas of savanna to thorn woodland in southern Texas probably result from scarification and dispersal of seeds by cattle (Archer, 1990). Climatic changes may increase erosion losses where low rainfall and/or cultural impacts have produced a rather discontinuous vegetation cover (Schlesinger et al., 1990; Walker, 1991). Erosion is worst where precipitation falls as high-intensity storms and where percolation is slow (i.e., where rainfall intensity exceeds infiltration rate). Because precipitation variability increases as average precipitation declines, frequency of flash floods may also increase with increased aridity. Arid and agricultural lands that support low plant cover much of the year with steep, shallow, and/or fine-textured soils are most sensitive. Numerous examples of changes in ecosystems as a result of these feedbacks include tropical grasslands of East Africa (Dublin et al., 1990; Belsky, 1992), fynbos shrublands in South Africa (Van Wilgren, 1980), submontane woodlands of Hawaii (D'Antonio and Vitousek, 1992), perennial grasslands and pinyon-juniper woodlands of the western United States, and grass/desert scrub of the American southwest (Schlesinger et al., 1990).

Fire. Fire depends on moisture availability via effects on fuel production, fuel-moisture content, and atmospheric conditions. Historically, fires were frequent in grasslands and woodlands. Deserts support discontinuous fuels that do not carry fires. Response of fire regime to climate change depends on short-term fluctuations that affect fuel moisture content and longer-term variability due to its effect on fuel production (Wright and Bailey, 1982). Wet years may permit high aboveground production that provides the fuels needed to support intense fires. High sensitivity of fire occurrence to climate variability is evident at the prairie-forest border, where temporal variations in weather conditions (Albertson and Weaver, 1945) and topographic variability and lake distributions (Grimm, 1984) have dramatic effects on vegetation as a consequence of fire.

Effects of climate change depend on an interaction between fire, N availability, and grazing. N fertilization has its greatest effects on productivity on recently burned sites (Seastedt et al., 1991). Reduced leaf area following fire allows for increased light at the soil surface and higher soil temperatures. Grazing reduces fuel loads and therefore the losses of N during fire (Hobbs et al., 1991). Grazing also increases the heterogeneity of vegetation (Schlesinger et al., 1990; Hobbs et al., 1991), therefore influencing fire behavior. Conversely, fire can serve to homogenize landscapes that have been affected by grazing (Hobbs et al., 1991). A given change in aridity, for example, will affect competitive and trophic relationships and fire regime, each of which experiences feedback effects via interactions with the other attributes of the systems. Nonlinear responses are likely in such circumstances (e.g., Schlesinger et al., 1990).

Grasslands are characterized by a continuous cover of herbaceous perennial grasses (Graminae) and grasslike species with two types of photosynthetic pathways. In the C_3 photosynthetic pathway, CO_2 is initially incorporated into a three-carbon compound. Most plants fix atmospheric carbon via the C_3 pathway. In the C_4 pathway, CO_2 is initially fixed into a four-carbon compound, which allows concentration of internal CO_2 for further fixation via the regular C_3 pathway. When high irradiance, high leaf temperature, and low transpiration reduce photosynthesis via internal CO_2 limitation, the C_4 pathway is more advantageous than C_3. Physiological differences in the two pathways confer a greater advantage on C_4 grasses in arid grasslands. The balance between species having C_3- and C_4-type photosynthetic pathways in grasslands is also affected by fertility and fire. The addition of N to tallgrass prairie results in a shift from the relatively nutrient-use-efficient C_4 grasses to increased representation of forbs (C_3) (Seastedt et al., 1991). With frequent fire, C_4 grasses may dominate, in part due to the low N availability on such sites. By increasing performance of the invasive C_3 annual grass *Bromus tectorum*, CO_2 enrichment could accelerate ecosystem transitions that attend increased fire occurrence with *Bromus* expansion in the Great Basin (Smith et al., 1987).

CO_2 Sensitivity

Rising atmospheric CO_2 could have important direct effects on species composition in grassland ecosystems, in part due to the mixture of C_3 and C_4 species that occur there. In temperate regions, where C_3 and C_4 grasses co-occur, the drought-intolerant C_3 species tend to dominate in spring and early summer, and chilling-sensitive C_4 species dominate in the drier late summer and autumn. Thus, while increasing water stress would favor earlier development of C_4 species, rising CO_2 would facilitate C_3 photosynthesis on into the summer months. Although C_3 is preferentially enhanced by elevated CO_2 in cool temperatures, field studies suggest that higher temperature favors C_4 grasses, which can shade out C_3 grasses (Nie et al., 1992). Elevated temperature and CO_2 also ameliorated growth responses of C_4 species for shortgrass prairies (Strain and Thomas, 1992). The combination of elevated CO_2 and climate change could increase competitive interactions among C_3 and C_4 grasses (Long and Hutchin, 1991).

Dry environments may see increases in productivity due to higher CO_2 exchange and higher water-use efficiency. Such a result would tend to increase water use because of a rise in total leaf area. But even qualitative predictions are difficult, because higher CO_2 might offset the effects of drought stress forecast by GCMs for many arid areas (Long and Hutchin, 1991). The photosynthetic capacity of C_3 trees will be increased by rising CO2, while that of C_4 grasses will be relatively unaffected. Expansion of *Prosopis* and *Quercus* (C_3 species) in C_4 grasses at the woodland-grassland boundary of southeastern Arizona may be partly due to CO_2 fertilization since the preindustrial period (McPherson et al., 1993). If water stress increases, however, the higher water-use efficiency of C_4 grasses at all CO_2 levels may prove more important (Long and Hutchin, 1991).

Salt Marshes

Salt marshes dominate trailing continental margins at temperate and high latitudes. They are generally productive systems that provide important habitat and breeding ground for diverse bird and invertebrate populations. Salt marshes are characterized by sedge and C_3 and C_4 grasses that are sensitive to salinity gradients with tidal influence and, therefore, elevation. C_3 plants dominate low elevations, where their deep root systems enable them to reach less saline water. C_4 species dominate upper marshes because of their greater drought and salt tolerance.

Salt marshes are likely to respond most dramatically to global changes in sea level and CO_2. Sea level affects soil properties through tidal flooding, including salinity, sulfide concentrations, and redox potential. Changes in species composition with changes in sea level are well documented in paleoecological (Clark, 1986a, b, c) and historical data (Warren and Niering, 1993).

Long-term field experiments under elevated CO_2 suggest that mixed communities of C_3 and C_4 species in a temperate salt marsh are more responsive than are pure stands (Arp et al., 1993). In the mixed community, a strong C_3 response was maintained throughout a four-year study without evidence of downregulation of photosynthetic capacity (Arp and Drake, 1991). In a pure C_3 stand, however, an initial biomass increase observed in the first two years of fumigation may have increased shading, thus reducing further responses. Although leaf-level net assimilation was greater for C_3, canopy assimilation was actually lower with elevated CO_2 (Drake, 1992). In pure C_4 stands, little change in biomass and density was observed with elevated CO_2. However, species distribution changed, with a decrease in the dominant *Spartina* biomass relative to other C_4 species, possibly because of shading by dominant *Scirpus* (C_3). Improved salt and drought tolerance may explain the expansion of the dominant C_3 to higher elevations and declining *Spartina* (C_4) (Arp et al., 1993).

CONCLUSIONS

Summary of High Sensitivities

Existing evidence suggests some important differences among biomes in their potential response to changing climate and atmospheric chemistry. High latitudes, particularly arctic ecosystems, are likely to witness highly nonlinear responses to temperature increase. Direct effects of rising atmospheric CO_2 will probably be less important in the arctic than are rising temperatures that melt permafrost; increase depth of the active layer; alter hydrology, C balance, and nutrient cycles; and potentially enhance growth via enhanced nutrient availability (table 1). Sensitivity to reduced irradiance because of increased cloudiness may partly alleviate this enhanced productivity. Prolonged growing season and

Table 1. Arctic processes potentially most sensitive to changes in atmospheric chemistry and climate

Scenario	Mechanism	Potential response	Sensitivity
Increased temperature	↑soil temperatures	↑ active-layer depth	High
		↑ evapotranspiration	Uncertain
		↓ water table	Probably high
	↑decomposition rate	↑ C release from organic soils	High
		↑ nutrient availability	High
		↑ productivity	High
	↑ photosynthetic rates	↑ productivity	High where nutrient availability also increases
	↑ growing season length	↑ productivity: tree invasion of southern margin	Uncertain
Increased cloudiness	↓ photosynthetic rates	↓ decreased productivity	Uncertain

increasing importance of woody plants promise a host of changes in community-level phenomena that cannot be forecast with any precision. New species assemblages will depend on a particular set of climate combinations that may not occur anywhere in the existing arctic. In general, the arctic is probably the biome most sensitive to rising temperatures.

Boreal ecosystems are likewise particularly sensitive to changing climate, while CO_2 effects are likely, albeit harder to predict. The variety of climate and CO_2 effects and feedbacks through sometimes contrasting influences on production, decomposition, fire, and insect outbreaks suggest a host of potential responses (table 2). Temperature and CO_2 will likely have the more obvious direct effects, but precipitation may also be important due to its influence on fire and competitive relationships. Rising temperatures could have important consequences for soil processes that affect hydrology, C storage, nutrient cycling, and productivity. Sensitivity of soil processes is highest in areas of permafrost. Also sensitive at longer time scales is the southern boundary of boreal forest, where invasion by broadleaved deciduous tree species will eventually alter C and nutrient pools. Fire will accelerate responses to global change, and fire regime is itself sensitive to changing climate, particularly in the more xeric western regions. Indirect effects on herbivores are likely but unpredictable. Response to rising CO_2 is likely but difficult to predict. The balance between photosynthetic and respiration responses is not well understood. And these responses will be coupled with possible interactive effects of altered temperature, nutrient, and moisture availability.

Climate and CO_2 sensitivities appear more variable across temperate forests due to a combination of factors. Sensitivity to change in temperature is expected to be highest near low-temperature physiological limits and where water balance is strongly limited by soil moisture-holding capacity or evaporative demand (table 3). Low-temperature tissue damage and temperature effects on rate processes suggest that (i) northern population frontiers of temperate tree species are strongly tied to temperature directly and (ii) southern ones depend on temperature indirectly through its effect on competition. Minimum winter temperatures, rather than averages, are expected to be most critical at the high-latitude limit of temperate forest. If rising temperatures at this limit are attended by an increase in minimum winter temperatures, broad-leaved deciduous forest is likely to invade the southern portion of boreal forest. The rate of this transition is expected to depend largely on disturbance, which will promote regeneration of more temperate species. The northward extent to which this occurs is difficult to predict, because minimum winter temperatures occur episodically. Dispersal distance will also affect this rate (Davis, 1983; Woods and Davis, 1989). It is expected that southern boundaries will be more diffuse and more closely tied to local site factors (Larcher, 1982; Woodward, 1987). We therefore predict high sensitivity to temperature change along northern population boundaries.

Moisture sensitivities follow local and regional water balance. Highly sensitive areas include the Midwest, where effective precipitation is low; the southeastern coastal plain, where soils are coarse but the water table is generally high; and mountainous terrain (e.g., much of the West), where soils are generally shallow and precipitation low. Fire and possibly the feedbacks between nutrient availability and vegetation contribute to the high sensi-

Table 2. Boreal processes potentially most sensitive to changes in atmospheric chemistry and climate

Scenario	Mechanism	Potential response	Sensitivity
Increased temperature	↑ soil temperatures	↑ active-layer depth	High in permafrost areas
	↑ decomposition rate	↓ soil C	Probably high, particularly where hardwoods invade
		↑ N & P availability	Probably high
		↑ productivity	Highest where nutrients now most limiting
	↑ photosynthetic rates	↑ productivity	Possibly high where nutrient availability increases
		↑ latitude and elevation of tree line	Responses most rapid on sites disturbed by fire
	↑ minimum winter temperatures	Invasion by broad-leaved temperate tree species	Southern margin of boreal forest
	↑ growing season length	↑ productivity ↑ C storage in living biomass expansion of *Abies*	Increase with latitude
	↑ fire importance	↑ soil temperatures ↑ active-layer depth ↑ decomposition rate ↓ soil C ↑ N & P availability	Midwestern (more xeric) areas (e.g., Manitoba)
Increased precipitation	↓ fire importance	↓ heterogeneity at some scales	Sensitivity uncertain
Increased cloudiness	↓ photosynthetic rates	↓ productivity	Sensitivity uncertain
Increased CO_2	↓ photosynthetic rates	↓ productivity	Sensitivity uncertain

tivity of prairie-woodland environments to climate. An important implication of this hypothesis is that the sensitivity of the prairie-woodland border to climate change may have decreased with fire suppression in the 20th century.

Table 3. Temperate forest processes potentially sensitive to changes in atmospheric chemistry and climate

Scenario	Mechanism	Potential response	Sensitivity
Increased temperature or decreased precipitation	↓ decomposition rate	↑ accumulation of low-quality litter	Depends on moisture status
		↑ nutrient availability	Depends on moisture status
Increased temperature and precipitation	↑ photosynthetic rates ↑ decomposition rates	↑ productivity	High where nutrient availability also increases
	↑ minimum winter temperatures	species composition: • invasion broad-leaved evergreen in south	Southern margin of temperate forest
		• ↑ recruitment success	Near high-latitude or high-elevation range limits of species
Increased cloudiness	↓ photosynthetic rates	↓ decreased productivity	Uncertain
Increased CO_2	↑ growth rate	competitive interaction, species composition	Where competitive interaction related to moisture stress and shade tolerance
	↑ tissue C:N	↓ decomposition rates and altered interactions with herbivores	Uncertain due to interactions with other processes

The transition over time from boreal to temperate deciduous forest will be among the most dramatic and rapid consequences of climate change. Carbon losses from organic-matter pools and changes in rates of nutrient turnover would occur on decade scales. Northward expansion of temperate forests would lead to large losses of soil organic matter, in part due to increased temperature and to higher-quality litter (Post et al., 1982). Increased mineralization rates would likely increase the availability of N and bolster productivity.

Although different geographic regions and local areas within the temperate zone are sensitive to climate change for different reasons, average sensitivities are probably not so great as those forecast for higher-latitude biomes, particularly wet tundra. Sites where soils have adequate water-holding capacity and those distant from critical temperature limitations

Table 4. Arid-land processes potentially sensitive to changes in atmospheric chemistry and climate

Scenario	Mechanism	Potential response	Sensitivity
Increased temperature	↓ photosynthetic rates	↓ productivity species composition changes, including C_3/C_4: C_4 favored by higher temperature; C_3 respond most to CO_2 fertilization and increased water-use efficiency	Where moisture limiting Savannas: more annual, less perennial grasses Grasslands: northward extension of C_4 grasses Tropical savannas and temperate grasslands (mixtures of C_3 and C_4)
	↑ evapotranspiration	prolonged summer drought ↑ fire where sufficient fuels, ↓ fire where fuel production declines	Mediterranean systems Savannas: ↑ fire ↓ woodiness
Increased precipitation	decomposition rates changed	productivity affected	Semiarid and arid
	↑ photosynthetic rates	↑ productivity ↑ fire where now limited by fuel availability	Grassland ecotones Grasslands
	↑ decomposition rates	↑ N availability in short term ↓ N availability in long term? ↑ C:N of vegetation ↑ fuel load due to more biomass of high C:N	Grassland, savanna
Increased seasonality	↑ variability in moisture supply and demand	altered fire regime shifts in dominant functional groups depending on phenology, diversity	Desert streams, dry rangelands, and savannas
	↑ variability in resource supply and demand	↑ runoff altered species composition	Savanna, grassland, desert
Increased CO_2	↑ productivity ↓ transpiration	species composition: C_3 favored	Where there is a mix of C_3, C_4

may show comparatively little response to climate change. Elevated CO_2 is likely to have its greatest effects through species-specific sensitivities to CO_2 fertilization and water use. Differential effects on species of different successional statuses, canopy positions, and herbivore loads are all potential ways in which rising CO_2 could alter species composition.

Changes in climate and atmospheric CO_2 both can have important consequences for grassland/savanna/desert ecosystems. Productivity, belowground C, nutrient availability, and fire depend importantly on water balance and probably atmospheric CO_2 (table 4). A shift in water balance will have short- and long-term effects on nutrient cycles in grassland ecosystems. Parallel shifts in CO_2, and thus water use, suggest complex responses that will be further affected by changes in fire regime and species composition. Because many of these regions are heavily impacted by grazing and cultivation, it is difficult to make specific predictions for particular areas.

Caveats

Our analysis suggests that overall sensitivity to the changing atmosphere is high everywhere, with the possible exception of the core of the eastern deciduous forest. Temperature effects at high latitudes; moisture effects in the West, Midwest, and Southeast; and CO_2 effects in the boreal forests and grasslands are likely. Sea-level rise and CO_2 have important consequences for coastal productivity. Nowhere do existing data suggest a straightforward CO_2 fertilization effect of the sort forecast for fertilized and watered agricultural systems. Caution should be exercised in extrapolating results of enhanced productivity by "CO_2 fertilization" in managed crop systems. Natural ecosystems, which have not been manipulated for high yield under management, may not be as responsive to this atmospheric fertilization. In all cases we examine here, CO_2 and climate interactions and human intervention complicate our ability to predict the outcome of changing the physical environment. Although reconstruction of past vegetation and climate association is useful, it does not represent a good model of future potential scenarios unless the impact of human disturbance is incorporated. For example, human influence has limited species dispersal by creation of small disjunct natural habitats between developed areas. On the other hand, species dispersal via human vectors has increased dramatically in today's society, whether unintentionally or by deliberate introduction of exotic species. Given that each of these biomes is exploited for its natural resources and/or agriculture, the changing climate and vegetation will be felt by many facets of society.

REFERENCES

Abrams, M. D., and M. L. Scott. 1989. Disturbance-mediated accelerated succession in two Michigan forest types. *Forest Science* 35:42-49.

Agee, J. K., and J. Kertis. 1987. Forest types of the North Cascades National Park Service Complex. *Canadian Journal of Botany* 65:1520-1530.

Albertson, F. W., and J. E. Weaver. 1945. Injury and death or recovery of trees in prairie climate. *Ecological Monographs* 15:393-433.

Almendinger, J. C. 1990. The decline of soil organic matter, total-N, and available water capacity following the late-Holocene establishment of jack pine on sandy mollisols, north-central Minnesota. *Soil Science* 150:680-694.

Archer, S. 1990. Development and stability of grass/woody mosaics in a subtropical savanna parkland, Texas, U.S.A. *Journal of Biogeography* 17:453-462.

Arp, W. J., and B. G. Drake. 1991. Increased photosynthetic capacity of Scirpus olneyi after 4 years of exposure to elevated CO_2. *Plant, Cell, and Environment* 14:1003-1006

Arp, W. J., B. G. Drake, W. T. Pockman, P. S. Curtis, and D. F. Whigham. 1993. Interactions between C_3 and C_4 salt marsh plant species during four years of exposure to elevated CO_2. *Vegetatio* 104/105:133-143.

Arris, L. L., and P. S. Eagleson. 1989. Evidence for a physiological basis for the boreal-deciduous forest ecotone in North America. *Vegetatio* 82:55-58.

Austin, M. P., and O. B. Williams. 1988. Influence of climate and community composition on the population demography of pasture species in semi-arid Australia. *Vegetatio* 77:43-49.

Bazzaz, F. A. 1990. The response of natural ecosystems to the rising global CO_2 levels. *Annual Review of Ecology and Systematics* 21:167-196.

Bazzaz, F. A., J. S. Coleman, and S. R. Morse. 1990. Growth responses of seven major co-occurring tree species of the northeastern United States to elevated CO_2. *Canadian Journal of Forest Research* 20:1479-1484.

Beasley, R. S., and J. O. Klemmendson. 1980. Ecological relationships of bristlecone pine. *American Midland Naturalist* 104:242-252.

Beer, T., A. M. Gill, and P. H. R. Moore. 1988. Australian bushfire danger under changing climatic regimes. Pp. 421-427 in G. I. Pearman, ed., *Greenhouse: Planning for Climate Change*. E.J. Brill, Leiden, Netherlands.

Belsky, A. J. 1992. Effects of grazing, competition, disturbance and fire on species composition and diversity in grassland communities. *Journal of Vegetation Science* 3:187-200.

Billings, W. D., J. O. Luken, D. A. Mortensen, and K. M. Peterson. 1982. Arctic tundra: A source or sink for atmospheric carbon dioxide in a changing environment? *Oecologia* 53:7-11.

_____. 1983. Increasing atmospheric carbon dioxide: possible effects on arctic tundra. *Oecologia* 58:286-289.

Birks, H. J. B. 1989. Holocene isochrone maps and patterns of tree-spreading in the British Isles. *Journal of Biogeography* 16:503-540.

Bonan, G. B., and H. H. Shugart. 1989. Environmental factors and ecological processes in boreal forests. *Annual Review of Ecology and Systematics* 20:1-28.

Botkin, D. B., J. F. Janak, and J. R. Wallis. 1973. Estimating the effects of carbon fertilization on forest composition by ecosystem simulation. Pp. 328-344 in G. M. Woodwell and E. V. Pecan, eds., *Carbon and the Biosphere*. U.S. Atomic Energy Commission. Symposium Series 30. National Technical Information Service, Springfield, VA.

Boul, S. W., P. A. Sanchez, J. M. Kimble, and S. B. Weed. 1990. Predicted impact of climate warming on soil properties and use. Pp. 71-82 in B. A. Kimball, ed., *Impact of Carbon Dioxide, Trace Gases, and Climate Change on Global Agriculture*. American Society of Agronomy, Madison, WI.

Bradbury, I. K., and D. C. Malcolm. 1978. Dry matter accumulation by Picea sitchensis seedlings during winter. *Canadian Journal of Forest Research* 8:207-213.

Braun, E. L. 1950. *Deciduous Forests of Eastern North America*. McGraw-Hill, New York.

Brown, K., and K. O. Higginbotham. 1986. Effects of carbon dioxide enrichment and nitrogen supply on growth of boreal tree seedlings. *Tree Physiology* 2:223-232.

Bryson, R. A. 1966. Air masses, streamlines, and the boreal forest. *Geographical Bulletin* 8:228-269.

Buell, M. F., and W. E. Martin. 1961. Competition between maple-basswood and fir-spruce communities in Itasca Park, Minnesota. *Ecology* 42:428-429.

Bunce, J. A. 1992. Stomatal conductance, photosynthesis and respiration of temperate deciduous tree seedlings grown outdoors at an elevated concentration of carbon dioxide. *Plant, Cell, and Environment* 115:541-549.

Bunce, J. A., L. E. Miller, and B. F. Chabot. 1977. Competitive exploitation of soil water by five eastern North American tree species. *Botanical Gazette* 138:168-173.

Burke, I. C., T. G. F. Kittel, W. K. Lauenroth, P. Snook, C. M. Yonker, and W. J. Parton. 1991. Regional analysis of the Central Great Plains. *BioScience* 41:685-692.

Burke, M. J., L. V. Gusta, H. A. Quamme, C. J. Weiser, and P. H. Li. 1976. Freezing and injury in plants. *Annual Review of Plant Physiology* 76:507-528.

Chapin, F. S., III, and G. R. Shaver. 1985a. Arctic. Pp. 16-40 in B. F. Chabot and H. A. Mooney, eds., *Physiological Ecology of North American Plant Communities*. Chapman and Hall, New York.

_____. 1985b. Individualistic growth response of tundra plant species to environmental manipulations in the field. *Ecology* 66:564-576.

Clarholm, M., P. Budimir, T. Rosswall, B. Söderström, B. Sohlenius, H. Staff, and A. Wiren. 1981. Biological aspects of nitrogen mineralization in humus from a pine forest podsol incubated under different moisture and temperature conditions. *Oikos* 37:137-145.

Clark, J. S. 1986a. Coastal forest tree populations in a changing environment, southeastern Long Island, New York. *Ecological Monographs* 56:259-277.

_____. 1986b. Dynamism in the barrier-beach vegetation of Great South Beach, New York. *Ecological Monographs* 56:97-126.

_____. 1986c. Late Holocene vegetation and coastal processes at a Long Island tidal marsh. *Journal of Ecology* 74:561-578.

_____. 1989a. Ecological disturbance as a renewal process: theory and application to fire history. *Oikos* 56:17-30.

_____. 1989b. Effects of long-term water balances on fire regime, northwestern Minnesota. *Journal of Ecology* 77:989-1004.

_____. 1990. Fire and climate change during the last 750 years in northwestern Minnesota. *Ecological Monographs* 60:135-159.

_____. 1993. Paleoecological perspectives on modeling broad-scale responses to global change. Pp. 315-332 in P. M. Kareiva, J. G. Kingsolver, and R. B. Huey, eds., *Biotic Interactions and Global Change*. Sinauer, Sunderland, MA.

COHMAP. 1988. Climatic changes of the last 18,000 years: Observations and model simulations. *Science* 241:1043-1052.

Cook, E. R., and J. Cole. 1991. On predicting the response of forests in eastern North America to future climatic change. *Climatic Change* 19:271-282.

Currie, D. J. 1991. Energy and large-scale patterns of animal-species and plant-species richness. *American Naturalist* 137:27-49.

D'Antonio, C. M., and P. M. Vitousek. 1992. Biological invasions by exotic grasses, the grass fire cycle and global change. *Annual Review of Ecology and Systematics* 23:63-87.

D'Arrigo, R., G. C. Jacoby, and I. Y. Fung. 1987. Boreal forests and atmosphere-biosphere exchange of carbon dioxide. *Nature* 329:321-323.

Daubenmire, R. F. 1936. The "Big Woods" of Minnesota: Its structure and relation to climate, fire, and soil. *Ecological Monographs* 6:233-268.

Davis, M. B. 1981. Quaternary history and the stability of forest communities. Pp. 132-153 in D. C. West, H. H. Shugart, and D. B. Botkin, eds., *Forest Succession: Concepts and Application*. Springer-Verlag, New York.

_____. 1983. Quaternary history of deciduous forests of eastern North America and Europe. *Annals of Missouri Botanical Garden* 70:550-563.

Davis, M. B., and C. Zabinski. 1992. Changes in geographical range resulting from greenhouse warming: Effects on biodiversity in forests. Pp. 297-308 in R. L. Peters and T. E. Lovejoy, eds., *Global Warming and Biological Diversity*. Yale University Press, New Haven, CT.

Dexter, F., H. T. Banks, and T. Webb, III. 1987. Modeling Holocene changes in the location and abundance of beech populations in eastern North America. *Review of Palaeobotany and Palynology* 50:273-292.

Drake, B. G. 1992. A field study of the effects of elevated CO_2 on ecosystem processes in a Chesapeake Bay Wetland. *Australian Journal of Botany* 40:579-595.

Dublin, H. T., A. R. E. Sinclair, and J. McGlade. 1990. Elephants and fire as causes of multiple stable states in the Serengeti-Mara woodlands. *Journal of Animal Ecology* 59:1147-1164.

Eamus, D., and P. G. Jarvis. 1989. The direct effects of increase in the global atmospheric CO_2 concentration on natural and commercial temperate trees and forests. *Advances in Ecological Research* 19:1-55.

Emanuel, W. R., H. H. Shugart, and M. P. Stevenson. 1985. Climatic change and the broad-scale distribution of terrestrial ecosystem complexes. *Climatic Change* 7:29-43.

Englemark, O. 1987. *Forest Fire History and Successional Patterns in Muddus National Park, Northern Sweden.* Department of Forest Site Research, Swedish University of Agricultural Sciences.

Engstrom, D. R., and B. C. S. Hansen. 1985. Postglacial vegetational change and soil development in southeastern Labrador as inferred from pollen and chemical stratigraphy. *Canadian Journal of Botany* 63:543-561.

Engstrom, F. B., and D. H. Mann. 1991. Fire ecology of red pine (*Pinus resinosa*) in northern Vermont, USA. *Canadian Journal of Forest Research* 21:882-889.

Fajer, E. D., M. D. Bowers, and F. A. Bazzaz. 1991. Performance and allocation patterns of the perennial herb, Plantago lanceolata, in response to simulated herbivory and elevated CO_2 environments. *Oecologia* 87:37-42.

Flannagan, M. D., and J. B. Harrington. 1988. A study of the relation of meteorological variables to monthly provincial area burned by wildfire in Canada (1953-80). *Journal of Applied Meteorology* 27:441-452.

Foster, D. R. 1982. The history and pattern of fire in the boreal forest of southeastern Labrador. *Canadian Journal of Botany* 61:2459-2471.

_____. 1988. Species and stand response to catastrophic wind in central New England, U.S.A. *Journal of Ecology* 76:135-151.

Gajewski, K. 1988. Late Holocene climate changes in eastern North America estimated from pollen data. *Quaternary Research* 29:255-262.

Garren, K. H. 1943. Effects of fire on vegetation of the southeastern United States. *Botanical Review* 8:617-654.

Greller, A. M., D. C. Locke, V. Kilanowski, and G. E. Lotowycz. 1990. Changes in vegetation composition and soil acidity between 1922 and 1985 at a site on the North Shore of Long Island, New York. *Bulletin of the Torrey Botanical Club* 117:450-458.

Grimm, E. C. 1983. Chronology and dynamics of vegetation change in the prairie-woodland region of southern Minnesota, U.S.A. *New Phytologist* 93:311-350.

_____. 1984. Fire and other factors controlling the Big Woods vegetation of Minnesota in the mid-nineteenth century. *Ecological Monographs* 54:291-311.

Grulke, N. E., G. H. Riechers, W. C. Oechel, U. Hjelm, and C. Jaeger. 1990. Carbon balance in tussock tundra under ambient and elevated CO_2. *Oecologia* 83:485-494.

Haines, D. A., V. J. Johnson, and W. A. Main. 1975. *Wildfire Atlas of the Northeastern and North Central United States*. U.S. Department of Agriculture Forest Service General Technical Report NC-16, Washington, DC.

Haines, D. A., W. A. Main, and E. F. McNamara. 1978. *Forest Fires in Pennsylvania*. United States Department of Agriculture Forest Service Research Paper NC-158.

Harmon, M. 1982. Fire history of the westernmost portion of Great Smoky Mountains National Park. *Bulletin of the Torrey Botanical Club* 109:74-79.

Heinselman, M. L. 1973. Fire in the virgin forest of the Boundary Waters Canoe Area, Minnesota. *Quaternary Research* 3:329-382.

_____. 1981. Fire intensity and frequency as factors in the distribution and structure of northern ecosystems. Pp. 7-57 in H. A. Mooney, T. M. Bonnicksen, N. L. Christensen, J. E. Lotan, and W. A. Reiners (eds.). *Fire Regimes and Ecosystem Properties*. U.S. Department of Agriculture Forest Service General Technical Report GTR-WO-26, Washington, DC.

Higginbotham, K. O. 1983. Growth of white spruce (*Picea glauca* [Moench] Voss) in elevated carbon dioxide environments. *Agriculture and Forestry Bulletin* 6:31-33.

Hobbs, N. T., D. S. Schimel, C. E. Owensby, D. S. Ojima. 1991. Fire and grazing in the tallgrass prairie: Contingent effects on nitrogen budgets. *Ecology* 72:1374-1382.

Host, G. E., K. S. Pregitzer, C. W. Ramm, D. P. Lusch, and D. T. Cleland. 1988. Variation in overstory biomass among glacial landforms and ecological land units in northwestern Lower Michigan. *Canadian Journal of Forest Research* 18:659-668.

Hunt, R., D. W. Hand, M. A. Hannah, and A. M. Neal. 1991. Response to CO_2 enrichment in 27 herbaceous species. *Functional Ecology* 5:410-421.

Insam, H. 1990. Are the soil microbial biomass and basal respiration governed by the climatic regime? *Soil Biology and Biochemistry* 22:525-532.

Jarvis, P. G. 1989. Atmospheric carbon dioxide and forests. *Philosophical Transactions of the Royal Society of London* Series B, 324:369-392.

Jarvis, P. G., and A. P. Sandford. 1986. Temperate forests. Pp. 199-236 in N. R. Baker and S. P. Long, eds., *Photosynthesis in Contrasting Environments*. Elsevier, Amsterdam, Netherlands.

Jenny, H. 1980. *Soil Genesis with Ecological Perspectives*. Springer-Verlag, New York.

Johnson, E. A. 1979. Fire recurrence in the subarctic and its implications for vegetation composition. *Canadian Journal of Botany* 57:1374-1379.

Kellogg, W. W., and Z.-C. Zhao. 1988. Sensitivity of soil moisture to doubling of carbon dioxide in climate model experiments. Part I: North America. *Journal of Climate* 1:348-366.

Knapp, A. K., and T. R. Seastedt. 1986. Detritus accumulation limits productivity of tallgrass prairie. *BioScience* 36:662-668.

Kullman, L. 1983. Past and present tree lines of different species in the Handolan Valley, central Sweden. Pp. 25-26. in P. Morisset and S. Payette, eds., *Tree Line Ecology*. Centre d'études nordiques de l'Université Laval, Quebec.

Kutzbach, J. E., and H. E. Wright. 1985. Simulation of the climate of 18,000 years BP; Results for the North American/North Atlantic/European sector and comparison with the geologic record of North America. *Quaternary Science Reviews* 4:147-187.

Lamb, H. F. 1980. Late Quaternary vegetational history of southeastern Labrador. *Arctic and Alpine Research* 12:117-135.

Lamb, H. F., and M. E. Edwards. 1988. The Arctic. Pp. 519-555 in B. Huntley and T. Webb III, eds., *Vegetation History*. Kluwer Academic, Leiden, Netherlands.

Larcher, W. 1982. Typology of freezing phenomena among vascular plants and evolutionary trends in forest acclimation. Pp. 417-426. in P. H. Li and A. Sakai, eds., *Plant Cold Hardiness and Freezing Stress*. Academic Press, New York.

Larcher, W., and H. Bauer. 1981. Ecological significance of resistance to low temperature. Pp. 403-437 in O. L. Lange, P. S. Nobel, C. B. Osmond, and H. Ziegler, eds., *Encyclopedia of Plant Physiology*. Springer-Verlag, Berlin, Germany.

Lincoln, D. E., and D. Couvet. 1989. The effect of carbon supply on allocation to allelochemicals and caterpillar consumption of peppermint. *Oecologia* 78:112-114.

Lincoln, D. E., D. Couvet, and N. Sionit. 1986. Response of an insect herbivore to host plants grown in carbon dioxide enriched atmospheres. *Oecologia* 69:556-560.

Long, S. P. 1991. Modification of the response of photosynthetic productivity to rising temperature by atmospheric CO_2 concentrations: Has its importance been underestimated? *Plant, Cell and Environment* 14:729-739.

Long, S. P., and P. R. Hutchin. 1991. Primary production in grasslands and coniferous forests with climate change: An overview. *Ecological Applications* 1:139-156.

Ludwig, D., D. D. Jones, and C. S. Holling. 1978. Qualitative analysis of insect outbreak systems: The spruce budworm and forest. *Journal of Animal Ecology* 47:315-332.

Lutz, H. J. 1930. Original forest composition in northwestern Pennsylvania as indicated by early land survey notes. *Journal of Forestry* 28:1098-1103.

Lutz, H. J., and A. L. McComb. 1935. Origin of white pine in virgin forest stands of northwestern Pennsylvania as indicated by stem and basal branch features. *Ecology* 16:252-256.

Luxmoore, R. J. 1981. CO_2 and Phytomass. *BioScience* 31:626.

MacArthur, R. H. 1972. *Geographical Ecology*. Princeton University Press, Princeton, NJ.

McAndrews, J. H. 1966. Postglacial history of prairie, savanna, and forest in northwestern Minnesota. *Memoirs of the Torrey Botanical Club* 22:1-72.

McPherson G. R., T. W. Boutton, and A. J. Midwood. 1993. Stable carbon isotope analysis of soil organic matter illustrates vegetation change at the grassland/woodland boundary in southeastern Arizona, USA. *Oecologia* 93:95-101.

Medina, E., and J. F. Silva. 1990. Savannas of northern South America—A steady state regulated by water-fire interactions on a background of low nutrient availability. *Journal of Biogeography* 17:403-413.

Meentenmeyer, V. 1978. Macroclimate and lignin control of litter decomposition rates. *Ecology* 59:465-472.

Menaut, J. C., J. Gignoux, C. Prado, and J. Clobert. 1990. Tree community dynamics in a humid savanna of the Côte d'Ivoire—Modeling the effects of fire and competition with grass and neighbors. *Journal of Biogeography* 17:471-481.

Minnich, R. A. 1983. Fire mosaics in southern California and northern Baja California. *Science* 219:1287-1294.

Mooney, H. A., B. G. Drake, R. J. Luxmoore, W. C. Oechel, and L. F. Pitelka. 1991. Predicting ecosystem responses to elevated CO_2 concentrations. *BioScience* 41:96-104.

Nie, D., M. B. Kirkham, L. K. Ballou, D. J. Lawlor, and E. T. Kanemasu. 1992. Changes in prairie vegetation under elevated carbon dioxide levels and two soil moisture regimes. *Journal of Vegetation Science* 3:673-678.

Norby, R. J., E. G. O'Neill, and R. J. Luxmoore. 1986a. Effects of atmospheric CO_2 enrichment in the growth and mineral nutrition of *Quercus alba* seedlings in nutrient-poor soil. *Plant Physiology* 82:83-89.

Norby, R. J., J. Pastor, and J. M. Melillo. 1986b. Carbon-nitrogen interactions inCO_2-enriched white oak: Physiological and long-term perspectives. *Tree Physiology* 2:233-241.

O'Neill, E. G., R. J. Luxmoore, and R. J. Norby. 1987. Elevated atmospheric CO_2 effects on seedling growth, nutrient uptake, and rhizosphere bacterial populations of *Liriodendron tulipifera* L. *Plant and Soil* 104:3-11.

Oechel, W. C., and W. T. Lawrence. 1985. Taiga. Pp. 66-94 in B. F. Chabot and H. A. Mooney, eds., *Physiological Ecology of North American Plant Communities*. Chapman and Hall, New York.

Oechel, W. C., and G. H. Riechers. 1986. Impacts of increasing CO_2 on natural vegetation, particularly the tundra. Pp. 36-42 in C. Rosenzweig and R. Dickinson, eds., *Climate-Vegetation Interactions*. Office for Interdisciplinary Earth Studies, University Corporation for Atmospheric Research, Boulder, CO.

Oechel W. C., G. Riechers, W. T. Lawrence, T. T. Prudhomme, N. Grulke, and S. J. Hasting. 1991. "CO_2LT": An automated, null-balance system for studying the effects of elevated CO_2 and global climate change on unmanaged ecosystems. *Functional Ecology* 6:86-100.

Oechel, W. C., and B. R. Strain. 1985. Native species responses to increased atmospheric carbon dioxide concentrations. Pp. 117-154 in B. R. Strain and J. D. Cure, eds., *Direct Effects of Increasing Carbon Dioxide on Vegetation*. DOE-ER-0238. U.S. Department of Energy, Carbon Dioxide Research Division, Washington, DC.

Overpeck, J. T., D. Rind, and R. Goldberg. 1990. Climate induced changes in forest disturbance and vegetation. *Nature* 343:51-53.

Parton, W. J., D. S. Schimel, C. V. Cole, and D. S. Ojima. 1987. Analysis of factors controlling soil organic matter levels in Great Plains grasslands. *Soil Science Society of America Journal* 51:1173-1179.

———. 1986. Influence of climate, soil moisture, and succession on forest carbon and nitrogen cycles. *Biogeochemistry* 2:3-27.

Pastor, J., and W. M. Post. 1988. Response of northern forests to CO_2-induced climate change. *Nature* 334:55-58.

Payette, S., and L. Filion. 1985. White spruce expansion at the tree line and recent climatic change. *Canadian Journal of Forest Research* 15:241-251.

Payette, S., L. Filion, L. Gauthier, and Y. Boutin. 1985. Secular climate change in old-growth tree-line vegetation of northern Quebec. *Nature* 315:135-138.

Payette, S., C. Morneau, L. Sirois, and M. Desponts. 1989. Recent fire history of the northern Quebec biomes. *Ecology* 70:656-673.

Piggott, C. D., and J. P. Huntley. 1981. Factors controlling the distributions of Tilia cordata at the northern limit of its geographical range. III. Nature and cause of seed sterility. *New Phytologist* 87:817-839.

Post, W. M., W. R. Emanuel, P. J. Zinke, and A. G. Strangenberger. 1982. Soil carbon pools and world life zones. *Nature* 298:156-159.

Prentice, I. C., P. J. Bartlein, and T. Webb, III. 1991. Vegetation and climate change in eastern North America since the last glacial maximum. *Ecology* 72:2038-2056.

Reid, C. D. 1990. The carbon balance of shade-tolerant seedlings of *Fagus grandifolia* and *Acer saccharum* under low irradiance and CO_2 enrichment. Ph.D. Dissertation, Duke University, Durham, NC.

Rind, D. 1988. The doubled CO_2 climate and the sensitivity of the modeled hydrologic cycle. *Journal of Geophysical Research* 93:538-541.

Ritchie, J. C., L. C. Cwynar, and R. W. Spear. 1983. Evidence from northwest Canada for an early Holocene Milankovitch thermal maximum. *Nature* 305:126-128

Sakai, A., and C. J. Weiser. 1973. Freezing resistance of trees in North America with reference to tree regions. *Ecology* 54:118-126.

Sala, O. E., W. J. Parton, L. A. Joyce, and W. K. Lauenroth. 1988. Primary production of the central grassland region of the United States. *Ecology* 69:40-45.

Sasek, T. W. and B. R. Strain. 1990. Implications of atmospheric CO_2 enrichment and climatic change for the geographical distribution of two introduced vines in the U.S.A. *Climatic Change* 16:31-51.

Schimel, D. S. 1988. Calculation of microbial growth efficiency from ^{15}N immobilization. *Biogeochemistry* 6:239-243.

Schimel, D. S., and W. J. Parton. 1986. Microclimatic controls of nitrogen mineralization and nitrification in shortgrass steppe soils. *Plant and Soil* 93:347-357.

Schimel, D. S., W. J. Parton, T. G. F. Kittel, D. S. Ojima, and C. V. Cole. 1990. Grassland biogeochemistry—links to atmospheric processes. *Climatic Change* 17:13-25.

Schlesinger, W. H. 1993. Response of the terrestrial biosphere to global climate change and human perturbation. *Vegetatio* 104-105:295-305.

Schlesinger, W. H., J. F. Reynolds, G. L. Cunningham, L. F. Huenneke, W. M. Jarrell, R. A. Virginia, and W. G. Whitford. 1990. Biological feedbacks in global desertification. *Science* 247:1043-1048.

Scholes, R. J. 1990. The influence of soil fertility on the ecology of southern Africa dry savannas. *Journal of Biogeography* 17:415-419.

Schulze, E. D., R. H. Robichaux, J. Grace, P. W. Rundel, and J. R. Ehleringer. 1987. Plant water balance. *BioScience* 37:30-37.

Seastedt, T. R., J. M. Briggs, and D. J. Gibson. 1991. Controls of nitrogen limitation in tallgrass prairie. *Oecologia* 87:72-79.

Sionit, N., B. R. Strain, H. Hellmers, G. H. Riechers, and C. H. Jaeger. 1985. Long-term atmospheric CO_2 enrichment affects the growth and development of *Liquidambar styraciflua* and *Pinus taeda* seedlings. *Canadian Journal of Forest Research* 15:468-471.

Smith, S. D., B. R. Strain, and T. D. Sharkey. 1987. Effects of CO_2 enrichment on four Great Basin grasses. *Functional Ecology* 1:139-143.

Solomon, A. M. 1986. Transient response of forests to CO_2 induced climate change: Simulation modeling experiment in eastern North America. *Oecologia* 68:567-579.

Sprugel, D. G. 1989. The relationship of evergreenness, crown architecture, and leaf size. *American Naturalist* 133:465-479.

Steijlen, I., and O. Zackrisson. 1987. Long-term regeneration dynamics and successional trends in a northern Swedish coniferous forest stand. *Canadian Journal of Botany* 65:839-848.

Stein, S. J. 1988. Explanations of the imbalanced age structure and scattered distribution of ponderosa pine within a high-elevation mixed coniferous forest. *Forest Ecology and Management* 25:139-153.

Stephenson, N. L. 1990. Climatic control of vegetation distribution: The role of the water balance. *American Naturalist* 135:649-670.

Strain, B. R., and F. A. Bazzaz. 1983. CO_2 and plants, Terrestrial plant communities. Pp. 177-222 in E. R. Lemon, ed., *CO_2 and Plants: The Response of Plants to Rising Levels of Atmospheric Carbon Dioxide*. Westview Press Inc., Boulder, CO.

Strain, B. R., and R. B. Thomas. 1992. Field measurements of CO_2 enhancement and climate change in natural vegetation. *Water, Air and Soil Pollution* 64:45-60.

Tieszen, L. L. 1975. CO_2 exchange in the Alaskan arctic tundra: Seasonal changes in the rate of photosynthesis of four species. *Photosynthetica* 9:376-390.

Tissue, D. T., and W. C. Oechel. 1987. Response of *Eriophorum vaginatum* to elevated CO_2 and temperature in the Alaskan tussock tundra. *Ecology* 68:401-410.

Tolley, L. C., and B. R. Strain. 1984a. Effects of CO_2 enrichment and water stress on growth of *Liquidambar styraciflua* and *Pinus taeda* seedlings. *Canadian Journal of Botany* 62:2135-2139.

_____. 1984b. Effects of CO_2 enrichment on growth of *Liquidambar styraciflua* and *Pinus taeda* seedlings grown under different irradiance levels. *Canadian Journal of Forest Research* 14:343-350.

_____. 1985. Effects of CO_2 enrichment and water stress on gas exchange of *Liquidambar styraciflua* and *Pinus taeda* seedlings grown under different irradiance levels. Oecologia 65:166-172.

Van Cleve, K., F. S. Chapin, III, C. T. Dyrness, and L. A. Viereck. 1991. Element cycling in taiga forests: state-factor control. *BioScience* 41:78-88.

Van Cleve, K., W. C. Oechel, and J. L. Hom. 1990. Response of black spruce (*Picea mariana* (Mill) B.S.P.) ecosystems to soil temperature modification in interior Alaska. *Canadian Journal of Forest Research* 20:1479-1489.

Van Wagner, C. E. 1978. Age-class distribution and the forest fire cycle. *Canadian Journal of Forest Research* 8:220-227.

Van Wilgren, B. W. 1980. Some effects of fire frequency on fynbos plant community composition and structure at Jonkershoek, Stellenbosch. *South African Forestry Journal* 18:42-55.

Vitousek, P. M. 1982. Nutrient cycling and nutrient use efficiency. *American Naturalist* 119:553-572.

Walker, B. H. 1991. Ecological consequences of atmospheric and climate change. *Climatic Change* 18:301-316.

Wardle, P. 1963. The regeneration gap of New Zealand gymnosperms. *New Zealand Journal of Botany* 1:301-315.

Waring, R. H. 1991. Responses of evergreen trees to multiple stresses. Pp. 371-390 in H. A. Mooney, W. E. Winner, and E. J. Pell (eds.) *Response of Plants to Multiple Stresses.* Academic Press, San Diego, CA.

Warren, R. S., and W. A. Niering. 1993. Vegetation change on a northeast tidal marsh: Interaction of sea-level rise and marsh accretion. *Ecology* 74:96-103.

Watts, W. A. 1970. The full-glacial vegetation of northwestern Georgia. *Ecology* 51:17-33.

_____. 1980. The late Quaternary vegetation history of the southeastern United States. *Annual Review of Ecology and Systematics* 11:387-409.

Watts, W. A., and B. C. S. Hansen. 1988. Environments of Florida in the late Wisconsin and Holocene. Pp. 307-323 in B. Purdy, ed., *Wet-Site Archaeology.* Telford Press, West Caldwell, NJ.

Webb, T., III. 1988. Eastern North America. Pp. 385-414 in B. Huntley and T. Webb III, eds., *Vegetation History.* Kluwer Academic, Leiden, Netherlands.

Webb, T., III, P. J. Bartlein, and J. F. Kutzbach. 1983. Holocene changes in the vegetation of the Midwest. Pp. 142-165 in H. E. Wright Jr., ed.. *Late Quaternary environments of the Eastern United States,* vol. 2, *The Holocene.* University of Minnesota Press, Minneapolis.

Whitney, G. G. 1986. Relation of Michigan's presettlement pine forests to substrate and disturbance history. *Ecology* 67:1548-1559.

Williams, W. E., K. Garbutt, F. A. Bazzaz, and P. M. Vitousek. 1986. The responses of plants to elevated CO_2 . IV. Two deciduous forest tree communities. *Oecologia* 69:454-459.

Woods, K. D., and M. B. Davis. 1989. Paleoecology of range limits: Beech in the upper peninsula of Michigan. *Ecology* 70:681-696.

Woodward, F. I. 1987. *Climate and Plant Distribution.* Cambridge University Press, Cambridge, UK.

Wright, H. A., and A. W. Bailey. 1982. *Fire Ecology.* Wiley, New York.

Zackrisson, O. 1977. Influence of forest fires on the North Swedish boreal forests. *Oikos* 29:22-32.

Zak, D.R., A. Hairston, D.F. Grigal. 1991. Topographic influences on nitrogen cycling within an upland pine oak ecosystem. *Forest Science* 37: 45-53.

5
The Importance of Nonlinearities in Global Warming Damage Costs

Stephen C. Peck[1] and Thomas J. Teisberg

The possibility of global warming due to greenhouse gas emissions presents a difficult policy problem because of the tremendous uncertainties involved. The cost of controlling greenhouse gases is uncertain, and the benefits are even more uncertain. One of the key uncertainties on the benefits side is the relationship between climate change and resulting damages (i.e., costs of impacts and of adaptations undertaken to reduce impacts).

In this paper, we examine the implications of uncertainty about warming damages, with a special focus on the role of nonlinearity in the damage function. After describing our analytic approach, we begin our analysis with a set of sensitivity cases to demonstrate the implications for optimal policy of alternative damage function assumptions. We find that optimal policy is much more sensitive to the degree of nonlinearity in damages than it is to the level of damages at a specified temperature increase.

Next, we present some simple value-of-information calculations designed to measure the benefits of accelerated resolution of uncertainties. First, we calculate the value of information about the level of damages and about the degree of nonlinearity of damages. Given the earlier sensitivity results, it is not surprising that we find that the value of information about damage function nonlinearity is higher than the value of information about the damage function level.

Finally, we explore the implications of damage function nonlinearity for the value of information about a key climate response parameter—the equilibrium temperature increase per carbon dioxide (CO_2) doubling, or "warming rate." We value information about the warming rate for three alternative maintained assumptions about the warming damage function—that it is linear, quadratic, or cubic. We find that the value of information about warming rate is much higher if the damage function exhibits highly nonlinear response to temperature change. Thus, the value of resolving other global warming uncertainties is shown to be importantly affected by the degree of nonlinearity in the damage function. These results underscore the central importance of research designed to determine the costs of impacts and adaptation, and in particular to determine the degree of nonlinearity in costs as a function of temperature change.

ANALYTIC APPROACH

The CETA Model

CETA (Carbon Emissions Trajectory Assessment) represents worldwide economic growth, energy consumption, energy technology choice, global warming, and global warming costs (costs of damage from and adaptation to higher temperature). Because CETA considers benefits of energy use together with costs of warming due to energy use, the model may be used to determine optimal time paths of CO_2 emissions and carbon taxes. These optimal paths reflect a balancing of the costs and benefits of emission reductions.

Structurally, CETA is closely related to Global 2100 (Manne and Richels, 1990, 1991a, 1992). However, CETA is an extension of Global 2100 insofar as it incorporates a model of global warming and a function representing the adaptation/damage costs resulting from warming. Also, CETA considers the warming problem from the perspective of the world as a whole, in contrast to the regional disaggregation of Global 2100.

Most of the data for the CETA model are adapted from the base case assumptions of the Stanford Energy Modeling Forum Global Climate Change study (Energy Modeling Forum [EMF], 1991).[2] In addition, a few key model parameters are either adapted from corresponding parameters used in Global 2100 or chosen to calibrate certain CETA outputs to corresponding Global 2100 outputs, for similar base case assumptions.

The following sections provide a brief overview of several of the key components of CETA. For a more detailed description of the model, see Peck and Teisberg (1992, 1993).

Energy Technologies

At the center of the CETA model is a production function which relates gross output to inputs of capital, labor, and energy. Energy inputs to the production function are determined endogenously from available energy technologies, which are classified as non-electric and electric. The non-electric technologies (and our shorthand abbreviations for them) are—

1. oil and gas (NE-OG),
2. direct coal use (COAL-D),
3. renewables (NE-RENEW),
4. synthetic fuels, or synfuels (SYNF),
5. a carbon-free backstop technology (NE-BACK);

and the electric technologies are—

1, 2, 3, 4. existing oil and gas, new oil and gas, existing coal, and new coal (reported together as E-FOSSIL);
5, 6. existing nuclear and hydroelectric (reported together as E-RENEW);
7. a carbon-free backstop technology (E-BACK).

Costs of energy use are related to energy use choices by unit production cost coefficients; these coefficients are shown in table 1, for each of the above energy technologies. In addition, use of each of the above technologies may be constrained in one or more ways, reflecting expectations about when new technologies first become available, expectations about the maximum reasonable scale for technologies over time, expectations about the phasing out of existing technologies, and exhaustible resource limits. Further, technology use may be subject to entry and exit constraints limiting the rate at which use may be phased in or phased out.

Table 1. Unit production cost coefficients

Non-electric technology	Cost ($/GJ)
Oil and gas	3.56
Direct coal use	2.00
Renewables	8.20
Synfuels	8.33
Non-electric backstop	16.67

Electric technology	Cost (mills/kWh)
Existing oil and gas	47.00
New oil and gas	56.50
Existing coal	23.40
New coal	51.00
Existing nuclear	20.60
Hydroelectric	2.80
Carbon-free backstop	50.00

Note: GJ = gigajoules = 10^9 joules; kWh = kilowatthours

Production Function

Besides energy, the other inputs to the production function are capital and labor. Capital input in the model is determined by an initial capital stock, endogenous investment in new capital, and the capital depreciation rate. Labor input is exogenously specified, and is expressed in efficiency units so that labor input reflects both the growth of the work force and improvement in work force productivity.

The actual production function in CETA is specified as a nested constant elasticity of substitution (CES) function. In this function, capital and labor inputs are combined in a Cobb-Douglas function as value-added input, and electric and non-electric energy inputs are combined in a Cobb-Douglas function as energy input. Value-added input and energy input then enter a CES function representing gross output. Mathematically, the production function is as follows:

$$Q = [A(K^{\alpha}L^{1-\alpha})^{\rho} + B(AEEI \cdot E^{\beta}N^{(1-\beta)})^{\rho}]^{1/\rho} \tag{1}$$

where

Q = output excluding energy sectors
K = capital input
L = labor input
E = electric energy input
N = non-electric energy input
A,B = scale factors
$AEEI$ = autonomous energy efficiency index
ρ = capital value share parameter
β = electric energy value share parameter
$ESUB$ = elasticity of substitution parameter
ρ = (ESUB-1)/ESUB

The autonomous energy efficiency index, AEEI, represents the potential for falling energy/ gross domestic product (GDP) ratios even at constant energy prices. This index is initialized at 1.0, and then rises over time at a rate of 1/4% per year.[3]

CO_2 Emissions

Using fossil fuel energy creates CO_2 emissions. In CETA, these are related to energy use by unit carbon emission coefficients; these coefficients are shown in table 2 for the energy technologies in CETA.

Warming

To relate CO_2 emissions from energy use to future global warming, CETA incorporates representations of the carbon cycle, the equilibrium temperature increase as a function of atmospheric CO_2 concentration, and the lag of actual temperature increase behind equilibrium increase.

The carbon cycle relates CO_2 emissions to future CO_2 concentrations. In CETA, we represent the carbon cycle using the simple response function estimated by Maier-Reimer and Hasselmann (1987). This response function summarizes the behavior of a coupled ocean-atmosphere carbon cycle model. The response function effectively divides new carbon emissions into five classes, each with a different atmospheric lifetime, ranging from infinity down to two years. Total atmospheric CO_2 over time is then the sum of atmospheric CO_2 across these five classes. The classes may be interpreted as representing ultimate repositories of carbon, including the atmosphere itself and four other nonatmospheric carbon sinks; each of the latter draws CO_2 from the atmosphere over a time scale that corresponds to the (finite) lifetime for that class.

Table 2. Unit carbon coefficients

Non-electric technology	Carbon coefficient (billion tons/EJ)
Oil and gas	0.0166
Direct coal use	0.0241
Renewables	0.0000
Synfuels	0.0400
Non-electric backstop	0.0000

Electric technology	Carbon coefficient (billion tons/tkWh)
Existing oil and gas	0.250
New oil and gas	0.133
Existing coal	0.323
New coal	0.253
Existing nuclear	0.000
Hydroelectric	0.000
Electric backstop	0.000

Note: EJ = exajoules = 10^{18} joules; tkWh = thousand kilowatthours.

The equilibrium temperature increase attributable to atmospheric CO_2 concentration depends on the logarithm of concentration (Dickinson and Cicerone, 1986) according to the following equation:

$$T_t = A\ln(C_t) - B \tag{2}$$

where

$\quad T_t$ = equilibrium temperature increase relative to preindustrial level, time t
$\quad C_t$ = CO2 concentration, time t
$\quad A,B$ = fitted parameters

The parameters A and B are fitted so that equilibrium temperature increase is zero at the preindustrial concentration, and 3 degrees Celsius (°C) at twice the preindustrial concentration. This reflects the current consensus estimate of the equilibrium increase for doubled CO_2 concentration (National Academy of Sciences, 1991). We refer to the equilibrium temperature increase per CO_2 doubling as the "warming rate"; it is a key climate response parameter in CETA.

In addition to equilibrium temperature increase from energy-related CO_2 emissions, CETA represents equilibrium increase from emissions of methane (CH_4), nitrous oxide (N_2O), and chlorofluorocarbons (CFCs), as well as from CO_2 emissions that result from land use changes. These emissions are exogenously specified at levels roughly consistent with Intergovernmental Panel on Climate Change (IPCC) scenario B, representing a moderate degree of future control (IPCC, 1990).[4] Atmospheric concentrations of CH_4, N_2O, and CFCs evolve over time according to a transition matrix (Nordhaus, 1990b) which governs the removal or transformation of atmospheric stocks of these gases over time. The equilibrium temperature increase attributable to these gases is derived from estimates in Wuebbles and Edmonds (1988) of the equilibrium temperature increase resulting from doubled concentrations of each. In total, all of these non-energy-related greenhouse gases make a small contribution to overall equilibrium temperature increase; during the 22d century, their projected contribution is about 15 to 17% of that attributable to energy-related CO_2.

Actual temperature increase lags the increase in equilibrium temperature. We use the following simple representation of this lagged adjustment process:

$$T_t^A = T_{t-1}^A + 0.02(T_{t-1} - T_{t-1}^A) \tag{3}$$

where

T_t^A = actual temperature increase, time t
T_t = equilibrium temperature increase, time t

The adjustment fraction, 0.02, is consistent with results from existing climate models (Schlesinger and Jiang, 1990).

Warming Adaptation/Damage Function

Costs of adaptation to and damage from global warming are extremely difficult to estimate. Nordhaus (1990a, b) has attempted to synthesize information from several sources to arrive at an estimate of the annualized cost to the United States of a 3°C warming.[5] In table 3, we present Nordhaus's estimates rearranged into two categories: costs in the national income accounts and costs not represented in these accounts. Costs in the national income accounts include reduced productivity in climate-dependent commercial activities, and land loss and shore protection costs due to sea level rise. Costs that are not counted include damage to unmanaged natural systems and changes in the amenity value of everyday life. The large number of question marks in the table reflects the large number of gaps in available information.

Climate-dependent commercial activities do not represent a large share of national income for the United States, and the effect of global warming on them is likely to be small and in some cases ambiguous or even positive in sign. For many activities, warming costs have not been estimated. Where estimates are available, costs tend to be small and contribute little to the overall total estimated cost of a 3°C warming.

Estimated losses due to sea-level rise are available and are the most significant warming costs. By far the largest cost due to sea-level rise is that of protecting high-value land from rising seas. This cost alone is about 85% of the total of quantifiable cost in the table.

Table 3. Impact estimates for different sectors for 3°C temperature increase

Sector	Billion 1981 $
Activities included in national income accounts	
Climate-dependent activities	
Agriculture	+12 to -12
Forestry, fisheries, other	small
Construction	positive
Water transportation	?
Air conditioning costs	+1.65
Reduced space heating costs	-1.16
Water and sanitary	negative
Hotels, lodging, recreation	?
Sea-level rise	
Land loss	0.48
Shore protection	5.70
Activities excluded from national income accounts	
Unmanaged natural systems	
Wetlands	?
Forests	?
Biological diversity	?
Amenities of everyday life	?
Total	
Central value for numerical estimates	6.67
Central value as % of national income	0.28

Source: Based on Nordhaus (1990a, b).

Perhaps the most important cost not represented in the national income accounts is damage to unmanaged natural systems. Such systems include wetlands and forests and the plant and animal species that live in them. The economic value of these systems primarily depends on the willingness of people to pay to preserve them from effects of warming. Undoubtedly, people are willing to pay something, and perhaps a substantial amount, to reduce impacts on these natural systems. However, estimates of willingness to pay are not available.

The total quantifiable cost in table 3 is about 1/4% of annual gross national product for the United States. Since this cost estimate omits important nonquantified costs, particularly damage to natural systems, and since costs could be higher in the less industrialized parts of the world, Nordhaus guesses that total costs for the world might be as high as 1 or at most 2% of gross production. We adopt the high end of this range as a benchmark cost for the damage functions we use here.

While Nordhaus's estimate is useful as a benchmark for the cost of a 3°C warming, we have no comparable benchmark for a higher temperature change, such as 6°C. If damages are linear in temperature change, the damage at 6°C would simply be twice the damage at 3°C. However, if damages are nonlinear increasing, as seems likely, damages at 6°C would be more than twice those at 3°C. As we will show later, it is very important to know the extent of the nonlinearity in the damage function.

For the analysis in this paper, we use damage functions which are power functions of the form:[6]

$$D_t = \alpha L_t \, (T_t^A)^\lambda \tag{4}$$

where

$$
\begin{aligned}
D_t &= \text{annual warming cost, time t} \\
\alpha &= \text{a scaling constant} \\
L_t &= \text{labor input index, time t (L1 = 1.0)} \\
T_t^A &= \text{temperature rise (above preindustrial)} \\
\lambda &= \text{power of the damage function}
\end{aligned}
$$

In our middle-case scenario, we set the damage function scaling constant, α, so that warming cost at 3°C is 2% of gross world production (GWP); this is consistent with our benchmark adopted from Nordhaus's work described above. Also, for our middle-case scenario, we set $\lambda = 2$, which implies a cost of 8% of GWP at a 6°C warming. Finally, we assume that the warming cost function shifts upward over time with growth in the labor index, L_t; this reflects the idea that willingness to pay to avoid warming effects will grow with population and income.

Solving CETA

CETA is specified as a nonlinear optimization problem.[7] The variables in this problem are investment and energy use choices (both amounts and technology types). The objective is to maximize the present-value utility of consumption, which is defined as gross output less investment, energy costs, and warming costs. The utility function is defined as population times the logarithm of consumption per capita,[8] and future utility is discounted at a rate of 3% per annum. We typically solve the nonlinear optimization problem for a time horizon extending out to the year 2250, using 10-year time intervals; however, we report results only out to the year 2200 to avoid time horizon end effects in the reported results.

Valuing Information with CETA

To estimate the value of information about a particular CETA parameter, we define possible values that the parameter might assume and probabilities associated with each of these values. Having done this, we estimate the value of information about a particular parameter as the difference between the expected present value obtainable if the true parameter value is known before the CO_2 control policy is chosen, less the expected present value obtainable if the CO_2 control policy must be chosen without knowledge of the parameter value. The first expected value is necessarily higher than the second, since knowledge of the true parameter value allows the CO_2 control policy to be precisely tailored to the actual value of the parameter.[9]

Since estimating the value of information with CETA is a computationally demanding exercise, we assume that only one parameter is uncertain at a time,[10] and we limit the number of possible values that a parameter might take to three: a low, middle, and high case. These three values are assigned probabilities 1/6, 2/3, and 1/6.[11] Then the CETA model is effectively specified as three parallel models, one for each "state of the world" (i.e., each possible value of the parameter being treated as uncertain).[12] This is done by creating a separate set of problem variables for each state of the world and a separate set of parallel problem constraints for each set of variables. Finally, the objective function is respecified as a probability weighted sum of the utilities obtained in each of the three parallel problems corresponding to the three states of the world.

With the model specified as described above, its solution is characterized by optimal variable values that are completely independent across states of the world, and by a maximized objective function value that is the expected utility when the parameter value is revealed before a CO_2 control policy must be chosen. Thus this model solution produces the expected utility obtainable with perfect information about the parameter value.[13]

The model specification described above can be further augmented with constraints that require variable values to be the same across states of the world for one or more time periods beginning with period 1. Then, for these time periods, the model determines a single set of variable values that maximizes expected utility across states of the world. The model solution in this case corresponds to the situation where the CO_2 control policy must be chosen before the parameter value becomes known. The difference between the expected present value obtained in this case and that obtained with perfect information then measures the expected value of obtaining perfect information now versus at some later date.[14]

OPTIMAL POLICY UNDER CERTAINTY

In this section, we explore the sensitivity of the optimal CO_2 emission control policy to variations in the parameters which define the warming adaptation/damage function. Our warming cost function is the power function described earlier and specified by two parameters: α, which scales the damage function, and λ, which is the power of the damage function. Our middle-case value for λ is 2, and our middle-case value for α is chosen so that costs for a 3°C warming would be 2% of GWP.

Middle-Case Damage Function Results

Figure 1 shows energy use over time for the middle-case damage function parameter values. Figure 1 reveals some key technology transitions that occur over the 200-year time horizon shown. In the electric sector, there is a transition from fossil fuel technologies (E-FOSSIL) to carbon-free backstop technology (E-BACK) around 2030. In the non-electric sector, two transitions occur: from oil and gas (NE-OG) to coal-derived synthetic fuels around 2050, another from coal-derived synfuels to NE backstop around 2140.

Figure 2 shows the time path of optimal carbon emissions for the middle-case damage function parameter values; for comparison, the figure also shows the time path of uncontrolled emissions. Note that emissions do not rise sharply until the transition from oil and gas to synthetic fuels occurs around 2050. Ultimately, the rise in emissions is curbed as the coal resource base supporting synthetic fuels becomes exhausted.[15] Note also that

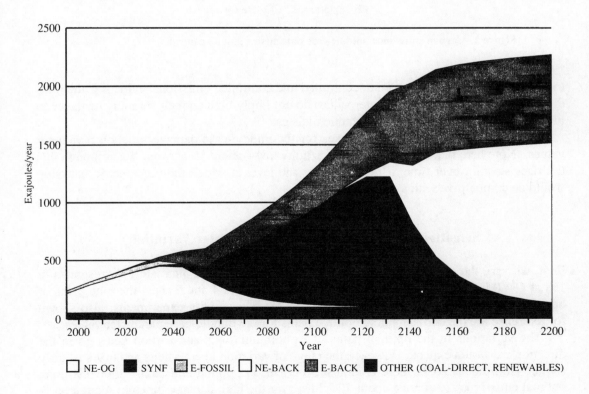

Figure 1. Energy use: middle-case parameters.

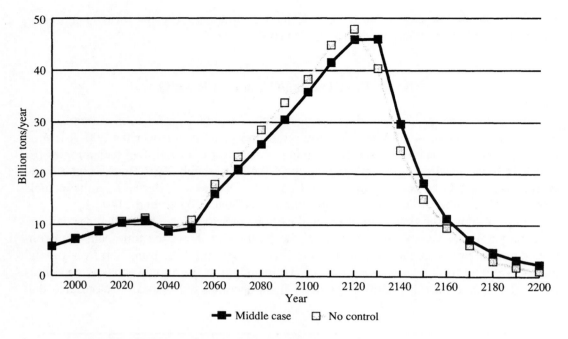

Figure 2. Carbon emissions: middle-case parameters and no control.

there is little difference between uncontrolled and controlled emissions; evidently, the middle-case damage function parameter values do not imply high enough warming damages to justify a significant reduction in synthetic fuels use.[16]

Figure 3 shows optimal carbon taxes for the middle-case damage function parameter values. Note here that the initial tax rate is quite low—about $9 per ton. Even though this tax rises steadily over time, it never reaches the level at which the non-electric backstop would be competitive with synthetic fuels.[17]

Sensitivity to Damage Function Parameter Variations

First, we vary the damage function scaling parameter so that it alternately corresponds to 1% of GWP in 1990 and to 3% of GWP in 1990. Figures 4 and 5 show the time paths of optimal carbon emissions and carbon taxes for these sensitivity experiments; also shown for reference are the corresponding optimal paths for the middle-case parameter values.

The variations in the optimal paths of carbon emissions and carbon taxes go in the direction we would expect. However, the range of variation in emissions and taxes is fairly small. Until about 2050, there is almost no effect on optimal emissions; after this time, optimal emissions are at most about 20% lower for the high damage function scaling parameter than they are for the low parameter.

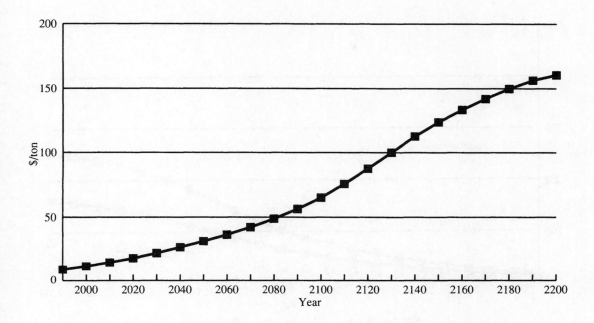

Figure 3. Carbon tax: middle-case parameters.

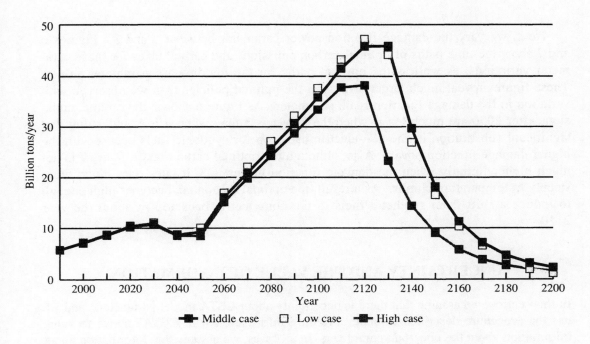

Figure 4. Carbon emissions: sensitivity to damage scale parameter.

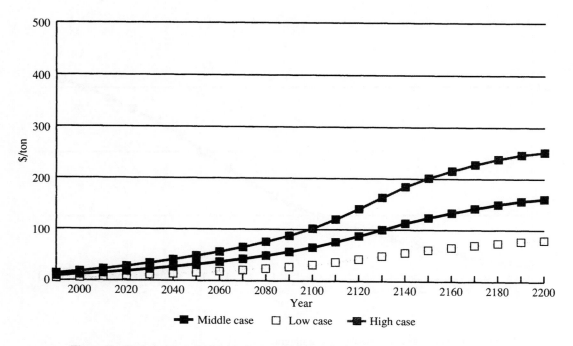

Figure 5. Carbon tax: sensitivity to damage scale parameter.

Next, we vary the damage function power parameter between 1 and 3. Figures 6 and 7 show the time paths of optimal carbon emissions and carbon taxes for these parameter variations, as well as the optimal paths for the middle-case parameter values. These figures reveal much larger swings in the optimal policies than we observed with variation in the damage function scale parameter. As figure 6 shows, the optimal emissions after 2050 are much lower when the power is 3 than when it is 1, indicating that significant substitution of the non-electric backstop for synthetic fuels occurs with the higher damage function power. Also, although the optimal carbon tax in figure 7 is not much higher initially when the damage function power is 3, it later rises much more steeply as temperature climbs. As a result, the optimal carbon tax becomes high enough to induce a shift from synthetic fuels to the non-electric backstop in about the year 2110.

UNCERTAINTY AND THE VALUE OF INFORMATION

In this section, we assume that there is uncertainty about CETA model parameters, and we use the procedure described in under "Valuing Information with CETA" above to value information about the uncertain parameters. In all cases, we assume that information about parameter values that is unavailable now will become available in 100 years. Thus our

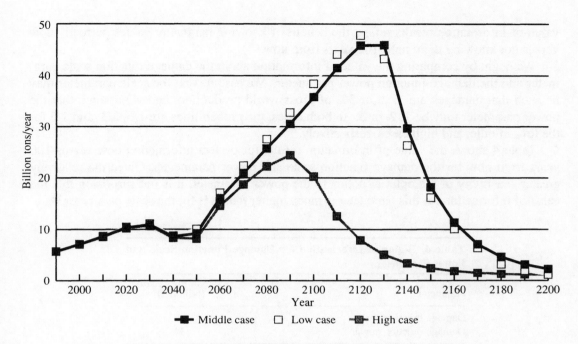

Figure 6. Carbon emissions: sensitivity to damage power parameter.

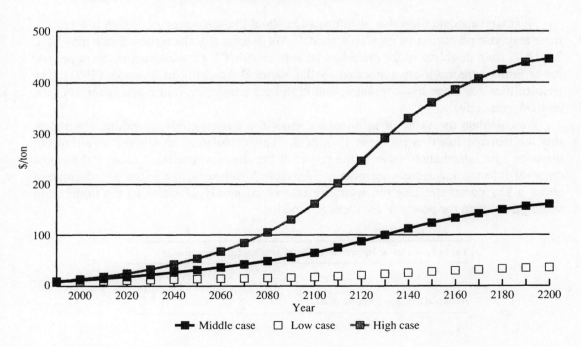

Figure 7. Carbon tax: sensitivity to damage power parameter.

value-of-information results reflect the benefit of knowing parameter values perfectly now versus not knowing them until 100 years from now.

We begin by comparing the value of information about the damage function scale parameter and the damage function power parameter. We assume that the scale parameter may be such that damages are 1, 2, or 3% of gross world production; and we assume that the power parameter may be 1, 2, or 3. In both cases, the probabilities are 1/6, 2/3, and 1/6 for the low, middle, and high cases, respectively.

Table 4 shows the value of information results for perfect information now versus 100 years from now for the damage function scale and power parameters. Given the relatively greater sensitivity of the optimal policy to the power parameter, it is not surprising that the value of information for this parameter is much higher than it is for the scale parameter.[18]

Table 4. Value of Information for Damage Function Scale and Power Parameters

Parameter	Value of information (billion $)
Damage function scale	6
Damage function power	34

Next we calculate the value of information about the warming rate, which is a key climate response parameter in the CETA model. We assume that the warming rate may be 1, 3, or 5°C for a doubling of the preindustrial atmospheric CO_2 concentration; the upper and lower limits here are those suggested by the National Academy of Sciences (1991). The probabilities for these low-, middle-, and high-case parameter values are again 1/6, 2/3, and 1/6, respectively.

We calculate the value of information about the warming rate, assuming alternately that the damage function power is 1, 2, or 3. This calculation reveals the sensitivity of warming rate information value to the power of the damage function. Table 5 shows the value-of-information results we obtain. As table 5 indicates, the value of information about a key parameter like the warming rate is extremely sensitive to the underlying assumption about the power of the damage function.

Table 5. Value of Information for Warming Rate Parameter

Assumed damage function power	Warming rate value of information (billion $)
1	3
2	88
3	665

CONCLUSIONS

In this paper, we have used the CETA model to explore the implications of nonlinearity in the adaptation/damage costs attributable to global warming. We find that optimal emissions control policies and carbon taxes are much more sensitive to the degree of nonlinearity in the damage function than they are to the scale of the damage function. We also find that the value of perfect information about the damage function power may be six times greater than the value of information about the damage function scale parameter. Finally, we find a strong interaction between the assumed damage function power parameter and the value of information we calculate for the warming rateæthis value of information is roughly two orders of magnitude greater if the damage function power is 3 rather than 1.

Based on these results, we conclude that the degree of nonlinearity in damages is a key uncertainty affecting optimal policy choice, as well as the benefits of resolving other global warming uncertainties. These results highlight the central importance of research designed to determine the costs of impacts and adaptation, and in particular to determine the degree of nonlinearity in costs as a function of temperature change. In general, we would argue that research agendas need to be guided by policy-oriented analyses like that we have presented here, rather than being driven solely by what are perceived to be the key scientific uncertainties surrounding a policy issue.[19]

NOTES

1. This paper does not represent the views of the Electric Power Research Institute or of its members.

2. The EMF study did not specify a total world coal resource base. We use an estimate obtained from Fulkerson et al. (1990). Their estimate is derived from World Energy Conference (1986); it includes proved reserves and the "Estimated Additional Amount in Place" which "might become recoverable within foreseeable economic and technological limits" (quote from World Energy Conference, 1986, notes to table 1.1).

3. The parameters ESUB and AEEI are similar to corresponding parameters in Global 2100. They allow energy growth and GDP growth to be decoupled. ESUB represents the potential for price-induced substitution of capital and labor for energy, while AEEI represents autonomous reductions in energy use.

   Since CETA has a putty-putty production function, while Global 2100 has a putty-clay function, output in CETA responds without lags to changing prices and autonomous energy efficiency improvements. Consequently, relative to Manne and Richels, we use somewhat lower values for ESUB (0.30 instead of 0.30 to 0.40 depending on region) and for AEEI (0.0025 instead of 0.005); the values we use are chosen so that key energy and GDP results from our model track those from Global 2100, for the same EMF input data.

4. The IPCC scenario extends as far as 2100. We assume that exogenous emissions continue to grow after 2100 at the same rate as prior to 2100.

5. The primary source Nordhaus relies on is U.S. Environmental Protection Agency (1989).

6. See Peck and Teisberg (1992) for a comparison of results obtained assuming that damages are related to the rate of warming versus the amount of warming.

7. We solve this problem using the GAMS/MINOS software package. See Manne (1986) for an introduction to the use of this software.

8. With constant population, this utility function would be equivalent to a simpler one, namely the logarithm of consumption. With growing population, however, the utility function we use tends to increase the marginal utility of consumption in the future relative to the marginal utility of consumption now, resulting in a slightly lower effective interest rate in the economy.

9. Our procedure here is the standard approach used to calculate the expected value of perfect information. See Raiffa (1968) for an introductory treatment.

10. It is much more difficult computationally to treat even two parameters as uncertain. However, in some preliminary experiments with two-parameter uncertainty, we have found that there is a synergistic interaction between uncertainties which causes the benefit of resolving any particular uncertainty to be higher if other parameters are assumed also to be uncertain.

11. We use an approach suggested by Miller and Rice (1983) for representing a normally distributed random variable with a three-point distribution. In this approach, the low and high cases given probability 1/6 correspond roughly to 5% and 95% probability points on the normal distribution being approximated; the middle case of course corresponds to the mean of the normal distribution.

12. We here follow the approach used by Manne and Richels (1991b) to value information using their Global 2100 model.

13. Expected present-value utility is converted to expected present-value dollars, using the shadow price of dollars in terms of utility. This allows values of information to be expressed in terms of dollars.

14. After the last time period in which variables are constrained to be the same, the model is once again free to tailor its solution to the actual state of the world. Thus parameter values are treated as uncertain only until the last date that variables are constrained to be the same.

15. Synthetic fuels may also be producible from shale, although our model does not reflect this possibility. If shale can be used, the available resources for making synthetic fuels are much larger, and the emission path would presumably turn down much later.

16. There is controversy about the appropriate rate at which to discount utility in an intergenerational analysis of this kind. We note that lowering the discount rate from 3 to 2% per year would cause optimal emissions to be reduced by about 7.8 billion tons per year in 2100 (about 20%).

17. Lowering the discount rate from 3 to 2% per year would cause the optimal carbon tax to approximately double to $136 per ton in 2100. (See note 16 in this paper.)

18. If the optimal policy is not sensitive to an uncertain parameter, there can be little benefit to knowing the true value of that parameter. Conversely, if the optimal policy is very sensitive to an uncertain parameter, there is likely to be high value to resolving uncertainty so that an appropriate policy can be adopted.

19. See Hidy and Peck (1991) for an elaboration of this viewpoint.

REFERENCES

Dickinson, Robert E., and Ralph J. Cicerone. 1986. "Future Global Warming from Atmospheric Trace Gases." *Nature,* vol. 319, no. 9 (January), pp. 109-115.

Energy Modeling Forum (EMF). 1991. "Study Design for EMF12 Global Climate Change: Energy Sector Impacts of Greenhouse Gas Emission Control Strategies." Draft (Stanford, CA: Stanford University, Energy Modeling Forum, May).

Fulkerson, William, Roddie R. Judkins, and Manoj K. Sanghvi. 1990. "Energy from Fossil Fuels." *Scientific American,* September, pp.128-135.

Hidy, G.M., and S.C. Peck. 1991. "Organizing for Risk Oriented Climate Alteration Research," *Journal of Air and Waste Management Association,* vol. 41, no. 12 (December), pp. 1570-1578.

Intergovernmental Panel on Climate Change (IPCC). 1990. *Climate Change: The IPCC Scientific Assessment,* Report prepared for IPCC by Working Group 1, Meteorological Office, Bracknell, UK.

Maier-Reimer, E., and K. Hasselmann. 1987. "Transport and Storage of CO_2 in the OceanæAn Inorganic Ocean-Circulation Carbon Cycle Model." *Climate Dynamics,* no. 2, pp. 63-90.

Manne, Alan S. 1986. "GAMS/MINOS: Three Examples." Draft (Stanford, CA: Stanford University, Department of Operations Research, December).

Manne, Alan S., and Richard G. Richels. 1990. "CO_2 Emission Limits: An Economic Cost Analysis for the USA." *The Energy Journal,* vol. 11, no. 2, pp. 51-74.

_____. 1991a. "Global CO_2 Emission Reductionsæthe Impacts of Rising Energy Costs." *The Energy Journal,* vol. 12, no. 1, pp. 87-107.

_____. 1991b. "Buying Greenhouse Insurance." *Energy Policy,* vol. 19, pp. 543-552.

_____. 1992. *Buying Greenhouse Insurance: The Economic Costs of CO_2 Emission Limits* (Cambridge, MA: MIT Press).

Miller, Allen C., III, and Thomas R. Rice. 1983. "Discrete Approximations of Probability Distributions." *Management Science,* vol. 29, no. 3, pp. 352-362.

National Academy of Sciences. 1991. *Policy Implications of Greenhouse Warming* (Washington, DC: National Academy Press).

Nordhaus, William D. 1990a. "To Slow or Not to Slow: The Economics of the Greenhouse Effect." Draft (New Haven, CT: Yale University, Department of Economics, February 5).

_____. 1990b. "Contribution of Different Greenhouse Gases to Global Warming: A New Technique for Measuring Impact." Draft (New Haven, CT: Yale University, Department of Economics, February 11).

Peck, Stephen C., and Thomas J. Teisberg. 1992. "CETA: A Model for Carbon Emissions Trajectory Assessment." *The Energy Journal,* vol. 13, no. 1, pp. 55-77.

_____. 1993. "Temperature Change Related Damage Functions: A Further Analysis with CETA." *Energy and Resources* (forthcoming).

Raiffa, Howard. 1968. *Decision Analysis: Introductory Lectures on Choices Under Uncertainty* (Reading, MA: Addison-Wesley).

Schlesinger, Michael E., and Xingjian Jiang. 1990. "Simple Model Representation of Atmosphere-Ocean GCMs and Estimation of the Time Scale of CO_2-Induced Climate Change." *Journal of Climate,* vol. 3, December, pp. 1297-1315.

U.S. Environmental Protection Agency. 1989. "The Potential Effects of Global Climate Change on the United States: Report to Congress, Executive Summary." Draft (Washington, DC: U.S. Environmental Protection Agency, December).

World Energy Conference. 1986. *1986 Survey of Energy Resources* (Oxford, UK: Holywell Press, Ltd.).

Wuebbles, Donald J., and Jae Edmonds. 1988. *A Primer on Greenhouse Gases*. DOE/NBB-0083. (Washington, DC: U.S. Department of Energy, March).

6

Sorting Out Facts and Uncertainties in Economic Response to the Physical Effects of Global Climate Change

Gary W. Yohe

Conversations between researchers who concentrate upon the effects of climate change along "best guess" future trajectories and people who worry about unforeseen surprises and nonlinearities are generally quite short and occasionally very unsatisfactory. Economic efficiency requires, for example, that the marginal cost of any policy designed to abate the pace of some global change phenomenon be weighed against its marginal bene-fitæthe marginal damage, net of adaptation but inclusive of the cost of that adaptation, that would otherwise be inflicted. But the author whose work quantifies the (marginal) benefit side of abatement by investigating adaptive responses along smooth, relatively likely trajectories can expect to face questions about how the calculus would change if some sort of sudden, surprising, and potentially discontinuous impact were felt somewhere along the way. Our author might respond by asking for a description of the surprise event to be considered, but that response would probably lead to the fairly incredulous retort that the event in question would not be a surprise if it could be described at the outset of the exercise. Even if it could be described hypothetically, our careful author would also want some idea of its relative likelihood. Only then could the damage that the surprise might inflict and the response that it might evoke be included in a subjective expected valuation of possible damages—the analysis required to bring the potential for surprise to bear in support of an elevated level of policy-driven abatement. The conversation might continue for a short while, in polite meetings anyway, but little headway should be expected.

The question which must be confronted is therefore one of incorporating what little we know about nonlinearities and surprises into the analysis which supports our decision making. If the methodology designed to overcome the impasse were sufficiently general, then there would be no need to be discouraged by how much is not known about what might happen. It would better inform current decision-making processes despite the paucity of data. It would allow new information consistently to be incorporated into future deliberations, and it could even provide insight into what type of new information would be most valuable.

This paper records some thoughts about how to proceed. It begins in the section entitled "A Taxonomy of Surprises" by recognizing the complication involved in understanding the scope of the natural and social processes which drive global change, but it continues by simplifying the portrait of the underlying natural and social systems in an effort to uncover a practical methodology. Focusing upon the possibility that strongly correlated, low-likelihood tails of distributions of effects might feed exaggerated impacts into nonlinear damage functions, it offers a straightforward approach to capturing the limits of "foreseeable surprises" which (1) meets the researchers' need for quantification and (2) recognizes the importance of including the damages associated with surprise events into economic analyses of abatement alternatives.

The next section, "Cascading Uncertainty and the Efficient Carbon Tax," gives the approach a try. The paper builds upon William Nordhaus's (1991) analysis of the efficient U.S. response to the threat of greenhouse warming along a "best guess" trajectory of damage, and computes an "uncertainty multiplier" for a wide range of other, less likely scenarios. We take the Intergovernmental Panel on Climate Change (IPCC) range of temperature sensitivity to a doubling of (effective) carbon concentrations as a measure of the uncertainty which surrounds current understanding of the physical phenomenon behind global warming. The effect of including uncertainty in this way is to multiply the Nordhaus estimate of the marginal cost associated with the efficient response for the United States to the effects of doubling by as little as 0.46 and as much as 15.00, depending upon the expected doubling temperature. The expected value of this uncertainty multiplier is 2.64, a value which increases the original Nordhaus estimate of the efficient reduction in cumulative carbon emissions for the United States through the year 2050 from 6% to nearly 15%.[1]

Building a method of systematically recognizing the exaggerated effect of the unfortunate coincidence of unlikely events and nonlinear damages clearly makes a difference. The precise numerical results recorded here are, of course, dependent upon the modeling and the interpretation of physical and economic data offered in their support. Some generalization of their significance, with particular reference to the broader issues of application, is offered under "Conclusions," but the implication of the actual multiplier computed here cannot be ignored. The effect of incorporating the potential cost of low-likelihood scenarios of what the future might hold is large enough to increase substantially the efficient response of the United States to the possible damage of greenhouse warming felt within her borders.

A TAXONOMY OF SURPRISES

Figure 1 displays a Social Process Diagram created during the summer of 1991 to complement the Physical Climate System "wiring diagram" depicted in figure 2.[2] The arrows which connect the box labeled "Global-Scale Environmental Processes" to the social, political, economic, and psychological structures scattered throughout figure 1 were, in fact, intended to mimic the arrows which link "Human Activities" on the right-hand side of

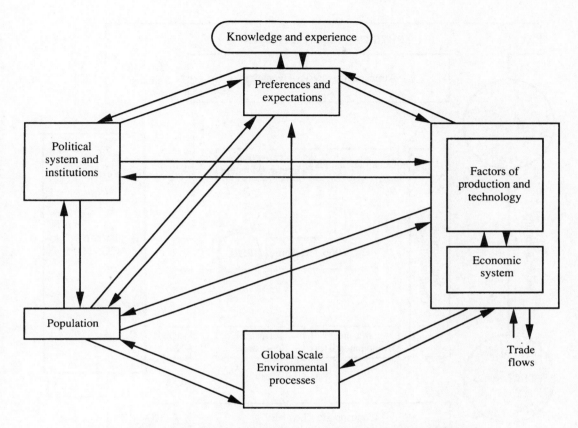

Figure 1. Social process diagram of the human dimensions of global change (CIESIN, 1992).

figure 2 with various aspects of the physical system. A common interface between social and physical processes is thus defined so that the two diagrams, taken together, illustrate the complicated interaction between human activity and the sources of global environmental change.

Economic, political, and social institutions are, of course, all linked in figure 1 even without reference to global environmental processes; their evolution over time should be driven by population and directed by preferences and expectations even in the absence of global-scale processes. Adding global physical processes to the picture simply adds another source of stress to the system, perhaps evoking decisions designed to ameliorate their effects, to slow their progress, or both. It is, however, impossible to judge the ability of human activity either to create global-scale processes or to influence their trajectories over time without a thorough understanding of the physical interactions of figure 2.

Figure 3 withdraws from the detail of the first two diagrams to offer an overly simplistic schematic of the link between global environmental change phenomena and their asso-

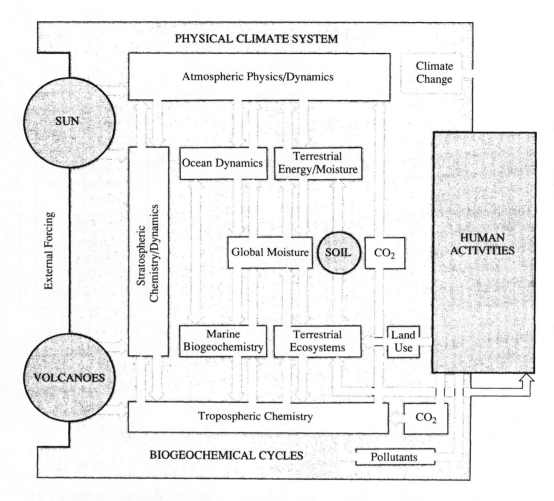

Figure 2. Physical systems schematic of global change. This wiring diagram, associated with Francis Bretherton, is adapted from the National Science Foundation publication MOSAIC (Fisher, 1988, pp. 56-57).

ciated economic damage. It is drawn to focus attention on the critical interface between figures 1 and 2. Despite masking the complication which clouds our understanding of the underlying physical and human processes of figures 1 and 2, its stark simplicity underscores vividly the compounding of processes that must be recognized in careful consideration of the full range of possible futures. Uncoupling this compounding process provides useful insight into how sources of significant surprise might be evaluated even given current, subjective, time-dependent views of how various component parts of the future might unfoldæviews which attribute very little likelihood to the outlying tails of the underlying distributions.

The first link in figure 3, labeled A, is meant to capture the workings of the entire Physical Climate System diagram. Something drives a change in the climate (e.g., a warming driven by increased concentrations of greenhouse gases), and a vector of physical impacts is expected. Certain impacts may be larger or smaller than others, but their relative importance to human activity cannot be judged in isolation. The second link, labeled B, works toward uncovering a reasonable basis upon which to make such a judgment. It reflects a computation of the resulting economic damage, but it is a computation which can be calibrated only after the ramifications of the impacts vector have been fully considered. Each impact must, in fact, work its way through the maze of the Social Process Diagram to include any adaptive and/or ameliorating response which might be forthcoming.

The enormous complexity of figures 1 and 2 notwithstanding, figure 3 can provide a convenient context for exploring the compounding potential of links A and B. Schedule I in figure 4 offers a straightforward portrait of a typical "best guess" scenario. It is the result of feeding a smooth, linear progression of physical change and associated impact over time for link A into a slowly increasing, linear correspondence between physical impact and economic damage for link B.[3] Schedule I therefore depicts a smooth, gradual expansion of economic damage over time, a picture of the least troubling of the possible outcomes of global change. Notice, for future reference, that the economic damage schedule has been drawn arbitrarily so that it climbs from an indexed value of 0 in the year 1990 to an indexed value of 1 by the year 2050.[4]

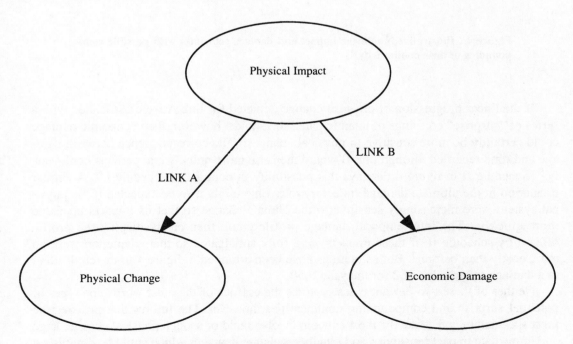

Figure 3. A simplistic representation of the integrated system of global change.

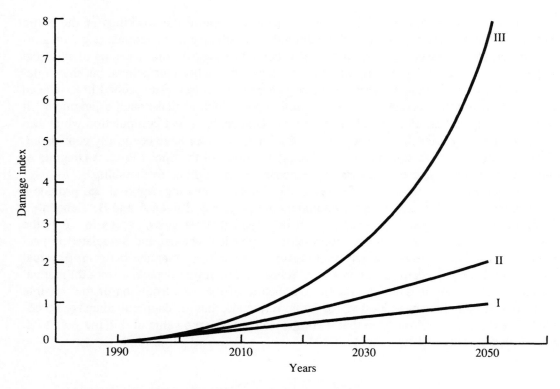

Figure 4. Illustrative, nonlinear impact and damage schedules with possible compounding of their nonlinearity.

If the linear progression of physical change depicted for link A were associated with a series of "surprise" crossings of unanticipated thresholds, however, then economic damage could certainly be more sensitive to physical changes. The correspondence between damage and time recorded through link B would then rise more quickly and perhaps nonlinearly.[5] Schedule II in figure 4 displays this possibility in contrast to schedule I.[6] A similar escalation in the ultimate damage trajectory over time might also be expected if the physical system were increasingly sensitive to the climate change so that its impacts appeared more quickly. The intertemporal damage profile could then easily display the profile offered by schedule II, if these impacts were fully anticipated so that adaptation were no more costly than before.[7] Both situations have been arranged in figure 4 to match, leading to a damage index value of 2 for the year 2050.

Neither of these two deviant cases captures the essence of the most worrisome types of potential surprise and compounding nonlinearities, however. The futures that provoke the most speculation and evoke the most concern involve rapid or sudden physical change, large and immediate impact sensitivity, <u>and</u> rapidly escalating damages which could be diminished only slightly, if at all, by adaptive response. Schedule III in figure 4 reflects such a possibility; its exaggerated discrepancy from schedule I is the result of compounding a nonlinear intertemporal impact trajectory.[8] Schedule II was drawn so that taking either source of non-linearity alone would create a 100% increase in the cost index by the year 2050. The nonlinearities displayed there combine to produce a damage index of 8 for 2050—an increase of 700%. The compounding effect shown along schedule III is thus the result of an unfortunate coincidence of two "bad" circumstances amplified by the nonlinearities which define each.

The shapes of the curves drawn in figure 4 are critical; and their precise shapes are, of course, entirely arbitrary. They can, nonetheless, be used to suggest one means by which the potential for significant surprises might be reflected in an appropriate deliberation of how to respond now to the possibility of global environmental change sometime in the future. Much of the best current debate is informed by subjective expected-value calculations, but little attention is typically paid to how the tails of the underlying distributions may coincide. Schedule III suggests, however, that even a cursory consideration of the interface between the physical impacts and the economic damages heralds the need to recognize the potential for strongly correlated series of "bad" circumstances. Its fundamental message is that a cascading confluence of tails of existing subjective distributions thrust against nonlinear impact and/or damage schedules can identify "surprises" whose relative likelihood and potential damage can both be quantified.

This sort of confluence can certainly suggest interesting "what if " scenarios to be considered, but it can, perhaps, do much more. Incorporating correlated surprises into the expected-value computations which frame the current policy debate could change both the relevant mean estimates of damages and costs and distributions of the appropriate responses. Unless the ultimate distributions are symmetric and the associated damage functions are linear, consideration of what happens along the "best guess" scenario is not the same thing as consideration of what happens along a certainty-equivalent scenario computed across a distribution of possible futures.

CASCADING UNCERTAINTY AND THE EFFICIENT CARBON TAX

Nordhaus (1991) offered the first comprehensive attempt to quantify the efficient response of any significant nation or region to the threat of greenhouse warming.[9] He recorded there the details of an initial effort to compute a response to the threat of greenhouse warming that would equate the marginal benefits of policies designed to reduce future concentrations of greenhouse gases with their marginal cost. His results, based upon baseline damage estimates drawn for the United States by the Environmental Protection Agency, attracted a considerable amount of attention—not only because they were firmly rooted in theoretically sound economic analysis, but also because they called for such a small response.[10] After expanding estimated damages by a factor of nearly 4 to capture unmeasured and unmeasurable impacts, in fact, the analysis suggested that efficiency criteria applied to midrange damages of roughly 1% of gross domestic product (GDP) supported only a 17% reduction in total greenhouse emissions engineered by growing a large number of trees, phasing out chlorofluorocarbon (CFC) consumption, and optimally reducing cumulative carbon emissions through 2050 by 6%. The marginal cost of this response, the shadow price of the targeted emissions reduction and thus the corresponding tax to be applied to carbon emissions, is roughly $13 per ton of carbon.

The Nordhaus work was, of course, based upon a set of "best guess" damage estimates calculated under the assumption that an effective doubling of carbon dioxide concentrations would occur around the year 2050. The major lesson of the previous section of this paper is, however, that looking at the tails of distributions of possible impacts and potential damages could easily substantiate warnings of exaggerated damages and thereby push estimates of expected damages well above "best guess" trajectories. The remainder of this section will employ the original Nordhaus analytical structure to explore the degree to which elevated estimates of damages that are consistent with current and coincident subjective views of future circumstances for the United States might enlarge the theoretically justified efficient response. It will, in other words, explore the effects of recognizing the potential for nonlinear damage functions and "foreseeable" surprises on the need to respond more vigorously to the threats of global change. Does, in short, the insight offered in the previous section matter at all?

A quick review of the structure of the Nordhaus model will be followed in turn by an extension which identifies an "uncertainty multiplier"—an index which reflects the degree to which the "best guess" scenario employed by Nordhaus to compute marginal damages causes him to underestimate (or, perhaps, to overestimate) the marginal damage associated with other possible futures. A review of the existing literature provides enough information about the structural components of that multiplier to support an informed attempt to quantify a subjective density function over its range and to judge its expected value. A final subsection will relate the associated distribution of uncertainty multipliers to a set of efficient marginal cost (carbon tax) statistics that apply across a plausible collection of possible futures. It is these statistics which are finally employed to equate the expected marginal benefit of reduced greenhouse gas concentrations (damage avoided) with the marginal cost of achieving that reduction.[11]

The Nordhaus Model

The operative Nordhaus model[12] began with a simplified temperature adjustment process characterized by:

$$(dT/dt) = a\{\mu M(t) - T(t)\}, \text{ with} \tag{1}$$

$$(dM/dt) = bE(t) - \delta M(t). \tag{2}$$

Notationally, the variables T(t), M(t), and E(t) represent the driving forces behind potential global environmental change. More specifically,

(i) $T(t)$ represents the increase in global mean temperature through time t generated by greenhouse warming since the preindustrial period of the middle of the last century;

(ii) $M(t)$ represents the atmospheric concentration of greenhouse gases at time t denominated in terms of carbon dioxide equivalents; and

(iii) $E(t)$ represents the emission in time t of greenhouse gases, again denominated in terms of carbon dioxide equivalents.

Parameters a, μ, b, and δ meanwhile define the relationships, with

(i) a reflecting a delay parameter which correlates a realized increase in temperature to a prior increase in radiative forcing;

(ii) b indicating the fraction of carbon-equivalent emissions which actually remain airborne;

(iii) δ representing a corresponding physical decay parameter for aggregated atmospheric concentrations of greenhouse gases; and

(iv) μ representing the (linearized) sensitivity of equilibrium temperature change to changes in atmospheric concentrations of greenhouse gases.

Equations (1) and (2) fully describe the link A structure required to operate within the simplistic schematic of figure 3.

The economic side of the model—link B in the parlance of figure 3—was meanwhile summarized by

$$c(t) = y(t)\{g(E^*) - \phi(T^*)\}, \text{ with} \qquad (3)$$

$$y(t) = y^* e^{ht}. \qquad (4)$$

Notationally,

(i) $c(t)$ represents per capita consumption at time t;

(ii) $y(t)$ represents per capita potential output growing in the absence of any emissions reduction and any deleterious effects of climate change at an annual rate of h;

(iii) $g(E^*)$ represents a steady-state computation of the cost of reducing emissions of greenhouse cases; and

(iv) $\phi(T^*)$ represents a steady-state computation of the economic damage associated with climate change.

It should be clear that temperature is being used as an index of climate change. Potential GDP per capita is reflected through equation (4). Equation (3) shows that consumption (per capita) can be diminished over time by the cost of abatement policies [the $g(E^*)$ term] and/or the damages caused by warming [the $\phi(T^*)$ term]. It should be noted with equal clarity that both costs are measured in steady state after any change in radiative forcing has achieved its long-run equilibrium.

The Nordhaus model operates as an exercise in long-run optimization, with

$$V = \int u[c(t)]e^{-\beta t}dt \qquad (5a)$$

serving as the objective function. The function $u[c(t)]$ is, of course, a utility function which allows for the possibility that the global society displays a systematic aversion to risk. For present purposes, however, including this possibility would "stack the deck" in favor of expanding uncertainty to include nonlinearities and surprises. It would hardly be news to conclude, under conditions of even moderate risk aversion, that surprises and non-linearities have important welfare effects which should be incorporated into the global decision-making calculus. It is, therefore, potentially far more interesting to see if recognizing the possibility of extreme events might have an effect even on risk-neutral decision making. If so, subsequent recognition of societal risk aversion would only exacerbate the effect. A risk-neutral objective function,

$$W = \int \{c[t]\}e^{-\beta t}dt \qquad (5b)$$

will thus be employed.[13] Abstracting to a risk-neutral objective function does not, of course, eliminate the need for specifying a real discount factor—the pure rate of time preference with which the present value of future consumption is computed. Risk neutrality does, however, imply that the elasticity of marginal utility with respect to per capital consumption is 0. The real rate of return on investment, denoted r, should therefore match the pure rate of time preference, the ß parameter in equations (5).[14]

The condition which characterizes the solution of the long-term optimization problem—maximize W subject to the constraints imposed on the system by equations (1) through (4)—is now at hand. It states quite simply that the present value of any small change in the emissions trajectory should be 0; i.e., the immediate increase in per capita consumption associated with a small increase in emissions should be matched by an increase in the present value of the damage, denominated in reduced consumption, associated with the long-run effect of those higher emissions. Nordhaus showed that this simple statement amounts to requiring that:

$$y^*g'(E^*)dE = \int [y^*e^{ht}\phi'(T^*)dT]^{-rt}dt \qquad (6a)$$

Equations (1) and (2) meanwhile combined under the assumption that $\delta << a$ to define $dT(t)$ in terms of physical parameters; more specifically,

$$dT(t) = \mu b e^{-\delta t}[1-e^{-at}]dE.$$

As a result, equation (6a) simplified immediately to

$$g'(E^*) = \mu b \phi'(T^*)A, \qquad (6b)$$

where the last term, A, is given by

$$A = \frac{1}{r - h + \delta} - \frac{1}{r - h + \delta + a}$$

It is equation (6b), under appropriate specification of the various physical and economic parameters, which supports the ultimate Nordhaus estimate of the efficient response to greenhouse warming, and it is an investigation of the effect of uncertainty and nonlinearities on the right-hand side of equation (6b) which will suggest the degree to which that efficient response should be adjusted to accommodate the possibility of surprises and non-linearities which can be quantified even now. An amended optimality condition—that the immediate increase in consumption associated with a small increase in emissions should be set equal to the present value of damage associated with the long-run effect of those emissions along any scenario—will produce a distribution of efficient responses contingent upon those scenarios. Aggregating over that range of response will then produce a second amended optimality condition—that the immediate increase in consumption associated with a small increase in emissions should be set equal to the _expected_ present value of damage associated with the long-run effect of those emissions.

Expected Marginal Damages of Emissions—An Uncertainty Multiplier

The marginal damage of emissions is the primary economic component of the right-hand side of equation (6b). As reported earlier, Nordhaus produced his point estimate of this component by relating damage statistics offered by the Environmental Protection Agency for a baseline warming scenario to the most vulnerable sectors of the national income accounts of the United States. He assumed, in creating his estimate, that the most likely scenario would see an effective doubling of carbon dioxide concentration by the middle of the next century. Generating a series of marginal damage estimates across a range of possible futures requires more than a single baseline estimate, however; it requires, instead, a marginal damage schedule defined throughout that range. Such schedules are few and far between, but one does exist for the economic vulnerability of the United States to greenhouse-induced sea-level rise.

Table 1 records estimates of national vulnerability for the United States as a function of greenhouse-induced sea-level rise presented initially in Yohe (1991a). The data recorded there are based upon a sample of over 30 sites systematically distributed along the entire coastline, so they take variation in the natural rate of subsidence into account. A simple econometric fit of their trajectory suggests that ψ(SLR), the economic vulnerability of the

United States coastline to sea-level rise (SLR), can be summarized by

$$\psi(SLR) = \psi_0 e^{g SLR}, \tag{7a}$$

with $g = 0.0253$ for SLR < 80 centimeters (cm) and $g = 0.0109$ for SLR > 80 cm.[15] Clearly, then,

$$\psi'(SLR) = g\psi_0 e^{g SLR} \tag{7b}$$

can be advanced as a possible candidate to serve as a proxy for the requisite marginal damage schedule.

Table 1. Economic vulnerability of the United States to greenhouse-induced sea-level rise

Sea-level rise[a]	Cumulative vulnerability[b]
13	36.3
21	46.1
30	68.6
40	110.7
53	137.2
60	153.9
67	191.1
83	270.5
100	308.7
132	424.3
165	701.7
200	909.4

[a] Sea-level rise induced by greenhouse warmingmeasured in centimeters.

[b] The vulnerability estimates are denominated in billions of 1989 dollars. Table 4 and table A.5 in Yohe (1991a) are their source; see the text of that paper for a description of the procedure employed in their creation.

The validity of using equation (7b) to represent overall marginal damages depends upon a number of considerations. The cost of sea-level rise along unprotected coastline did, however, sum with the cost of protection to quantify the dominant source of potential damage for the United States in the Nordhaus work. Advancing equation (7b) as a rough approximation of at least the proper form of the marginal damage component of equation (6b) is not totally unwarranted, at least not if one assumes that this dominance will persist over the range of possible futures and that the cost of protection should rise proportionately with the economic vulnerability of effected locations. Its structure is, moreover, certainly consistent with the notion that increasingly severe changes in climate should move the earth along nonlinear damage functions because they will be associated with increasingly frequent episodes of costly effects and adaptation.

Distributions of future sea-level rise related to anticipated increases in equilibrium temperature are, in addition, required to support the structure of equation (7b). Given any specific expectation about the temperature sensitivity of doubling, recent work by Wilson (1988), Oerlemans (1989), and Shlyakhter and Kammen (1992a, b) suggests that an exponential distribution of predicted sea-level rise is most appropriate. A density function

$$f(\text{SLR}) = \alpha e^{-\alpha(\text{SLR})} \qquad (8)$$

can thus be advanced with a mean $[= (1/\alpha)]$ dependent upon an assumed doubling sensitivity for temperature. The IPCC Scientific Assessment meanwhile offers a best guess that an effective doubling of carbon dioxide concentrations would force a 2.5 degrees Celsius (°C) increase in equilibrium temperature and a 66-cm increase in sea-level rise by the year 2100. The low end of the temperature range reported there stands at 1.5°C, presumably associated with the low end of the reported potential for sea-level rise (33 cm through 2100); the high end of temperature sensitivity stands at 4.5∞C with a 99-cm sea-level rise. Using $T_0 = 2.5$°C as the basis for a temperature index, $[T_d/T_0]$, and taking the IPCC sea-level scenarios as mean estimates, a two-part linear relationship between the temperature index and the required in the underlying sea-level density function was computed:

$$\alpha(Td) = 0.040 - 0.040[T_d/T_0] \quad [T_d/T_0] < 1$$
$$\alpha(Td) = 0.023 - 0.007[T_d/T_0] \quad [T_d/T_0] \geq 1. \qquad (9)$$

Equations (8) and (9) combine to relate distributions of sea-level rise to anticipated temperature increases associated with an effective doubling of carbon dioxide concentrations.

It is most convenient, at this point, to bring equations (7b), (8), and (9) together to synthesize a final representation of the expected marginal damage function—one fundamental component of the right-hand side of equation (6a). Let SLR_0 represent the expected sea-level rise associated in equilibrium with the IPCC best-guess doubling temperature sensitivity of 2.5°C. A short Taylor expansion of the right-hand side of equation (7b) can also be employed to produce

$$E\{\psi'(T_d)\} = \psi_0 g\text{SLR}_0 e^{g\text{SLR}_0} \int e^{g[\text{SLR}(T_d)-\text{SLR}_0)}f(\text{SLR}/T_d)d\text{SLR} \qquad (10a)$$

for any given T_d, with

$$f(\text{SLR}/T_d) = \alpha(T_d)e^{-\alpha}(T_d).$$

Performing the integration indicated allows a dramatic simplification of equation (10a):

$$E\{\psi'(T_d)\} = \psi_0 g \text{SLR}_0 e^{g\text{SLR}_0} \frac{\alpha(T_d)}{[\alpha(T_d) - g]} e^{-g\text{SLR}_0}. \tag{10b}$$

The cost structure thereby fully described, attention must now turn the other terms in the right-hand side of equation (6b) and their relationship with the distribution of anticipated doubling temperature increases.

The first term appearing there is μ, the sensitivity of increase in equilibrium temperature to a change in the concentration of greenhouse gases. When equilibrium has been achieved, of course, $[dT(t)/dt] = 0$, so equation (1) reduces to

$$T^*(t) = \mu M^*(t).$$

Recalling a commonly employed relationship between doubling temperature and concentrations,[16] specifically that

$$T(t) = \frac{T_d}{\ln (2)} \ln\{M(t)/M(0)\}$$

$$\approx \{T_d / [M(0)\ln(2)]\}\{M(t) - M(0)\},$$

it becomes reasonable to assert that

$$\mu \approx \{T_d / [M(0)\ln(2)]\} = \{T_0 / [M(0)\ln(2)]\}\{T_d/T_0\}.$$

It becomes essential, therefore, to explore the subjective distribution of the previously defined index of doubling temperature sensitivity—i.e., the distribution of $\{T_d/T_0\}$.

The IPCC Scientific Assessment (1990) provides some insight into that distribution, but not very much. It records, in table 3.2(a), a series of recent studies which report equilibrium doubling temperature increases between 1.9°C and 5.2°C. The authors weigh this evidence against modeling results provided to the IPCC from independent researchers to conclude that

> the sensitivity of global mean surface temperature to doubling (effective) carbon dioxide (concentrations) is unlikely to lie outside the range of 1.5 to 4.5°C. There is no compelling evidence to suggest in what part of this range the correct value is most likely to lie. There is no particular virtue in choosing the middle of the range, and both the sensitivity and the observational evidence neglecting factors other than the greenhouse effect indicate that a value in the lower part of the range may be more likely. Most scientists decline to give a single number, but for the purpose of illustrating IPCC scenarios, a value of 2.5°C is considered to be the "best guess" in the light of current knowledge [p. 139].

Setting 2.5°C as the benchmark "best guess" T_0, this passage suggests that the doubling temperature increase index $\{T_d/T_0\}$ could be as low as 0.6 or as high as 1.8. Placing equal weight on the likelihood that 0.6, 1.2, and 1.8 will turn out to be the correct value (to reflect the "no compelling evidence to suggest in what part of this range the correct value is most likely to lie" phrase) yields an implicit variance for a representative subjective distribution over index number range of 0.24. Imposing a gamma distribution over the range to capture the notion that the "best guess" index number is 1 (for a doubling temperature of 2.5°C) then suggests that $f_{T_d}\{T_0\} \equiv \Gamma(5,6)$ might be a reasonable density function. Table 2 displays the resulting relative frequency weights across the prescribed range of possible values, showing a modal value of 1 lying just below the median; figure 5 illustrates the truncated distribution across the temperature sensitivity interval [0.6, 1.8].[17]

Table 2. The relative frequency of the increase in global mean temperature associated with an effective doubling of carbon emissions

(1)a T_d/T_0	(2)b Relative frequency
0.6	6.0%
0.7	7.9
0.8	9.3
0.9	10.2
1.0	10.5
1.1	10.2
1.2	9.6
1.3	8.7
1.4	7.6
1.5	6.5
1.6	5.5
1.7	4.5
1.8	3.6

a An index of the increase in equilibrium global mean temperature associated with an effective doubling of atmospheric concentrations of carbon dioxide with 1 = 2.5°C.
b Relative frequency computed according to $\Gamma(5,6)$, truncated to run from 1°C to 4.5°C.

The other terms in the right-hand side of equation (6b) are less problematical and less important. The discount parameter A depends, for the most part, upon economic parameters which are best handled with sensitivity analysis. The one exception is δ, the atmospheric decay parameter, which is small relative to $(r - h)$ and negatively correlated with b, the airborne fraction parameter. The value assumed by this airborne fraction can be taken to be uncorrelated with the doubling sensitivity of the climate, since it effects a process which occurs before radiative forcing occurs. It is therefore unlikely that either b or A will systematically influence either the $\{T_d/T_0\}$ index or the marginal damages associated with a particular climate change effect, and uncertainty in both will be ignored.

It is now possible to express the expected value of the right-hand side of equation (6b) in terms which are familiar and easily interpretable. Allowing b, A, r, h, and δ now to represent either the expected values of underlying random variables which are uncorrelated with anticipated doubling sensitivities or specific values for exogenous economic variables which frame the overall growth context of the greenhouse problem, that expected value can now be written as

$$E\{RHS\} = Ab\,\frac{T_0}{\ln\{2\}}\int[T_d/T_0]E\{\psi'(T_d)\}f_{T_d}(T_d)dT_d$$

$$= Ab\,\frac{T_0}{\ln\{2\}}\psi_0 g SLR_0 e^{g SLR_0}\int[T_d/T_0]\frac{\alpha(T_d)}{[\alpha(T_d)-g]}\,e^{-g SLR_0}f_{T_d}(T_d)dT_d\,.$$

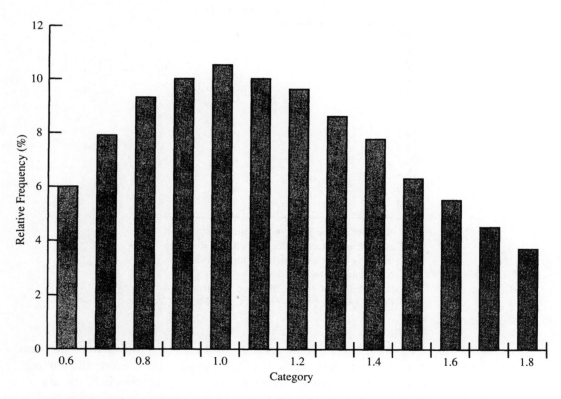

Figure 5. A graphical representation of the gamma density function for the doubling temperature change index ($T_0 = 2.5°C$).

Everything to the left of the integral sign is captured in the Nordhaus baseline estimate of marginal damage. Everything to the right, therefore, is an uncertainty index which can exaggerate or diminish the original baseline statistic. The key to producing an understanding of the degree to which uncertainty would cause the expected marginal damage of increased emissions to exceed the baseline estimate therefore lies in understanding the degree to which this uncertainty index,

$$\pi = \int [T_d/T_0] \frac{\alpha (T_d)}{[\alpha(T_d) - g]} e^{-gSLR_0} f_{T_d}(T_d) dT_d$$

$$\equiv \int [T_d/T_0] D[T_d] f_{T_d}(T_d) dT_d$$

$$\equiv \int \pi [T_d/T_0] D f_{T_d}(T_d) dT_d$$

exceeds unity. The key to producing a range of possible marginal damage statistics contingent upon specific temperature sensitivities within the quoted IPCC range similarly lies in investigating the range of values which might be assumed by the various $\pi[T_d/T_0]$—the weighted marginal damage multipliers defined implicitly by equation (11).

Quantifying the Uncertainty Multiplier

Table 3 displays the critical results, given the specifics of the modeling extension described above. The second column records the marginal damage multiplier for each $[T_d/T_0]$ index—the D[Td] parameter implicitly defined in equation (11) as

$$D[T_d] = \frac{\partial (T_d)}{[\alpha(T_d) - g]} e^{-gSLR_0} .$$

Column (3) records the corresponding weighted marginal damage multiplier—the $\pi[T_d/T_0]$ computed according to equation (11) as the product of the index value of column (1) and the damage multiplier shown in column (2). Notice that these uncertainty multipliers run from a low of 0.46 for a doubling temperature index of 0.6 (doubling associated with 1.5°C), to a high of 15.00 for a temperature index of 1.8 (an equilibrium doubling temperature of 4.5°C).

These values represent the expected damage multiplier contingent upon the indicated value of $[T_d/T_0]$. For $[T_d/T_0] = 0.6$, for example, $T_d = 1.5°C$ and $\pi[T_d/T_0] = 0.46$, so that the expected marginal damage estimate given a doubling temperature of 1.5°C is $5.84.[18] For $[T_d/T_0] = 1.0$, $T_d = 2.5°C$, $\pi[T_d/T_0] = 1.42$ is the resulting expected damage multiplier, and $18.03 represents the contingent estimate of expected marginal damage. Notice that uncertainty in our understanding of possible sea-level rise trajectories even given the best-guess temperature estimate increases marginal damage by 42%. On the opposite extreme, the contingent estimate of expected marginal damage is $190.50 when $T_d = 4.5°C$, so that $[T_d/T_0] = 1.8$, and the expected damage multiplier is 15.00. The expected value calculation prescribed by equation (11) yields a mean of 2.64 over the entire range—a value roughly matching the 70th percentile of the $\pi[T_d/T_0]$ distribution.

Table 3. The uncertainty multiplier, marginal damages, and efficient reductions in carbon emissions

(1)a T_d/T_0	(2)b $D[T_d]$	(3)c $\pi[T_d/T_0]$	(4)d Marginal Damage	(5)e Efficient Reduction
0.6	0.76	0.46	$5.84	2.8%
0.7	0.86	0.60	7.62	3.7
0.8	1.03	0.82	10.41	5.0
0.9	1.37	1.28	16.26	7.6
1.0	1.42	1.42	18.03	8.5
1.1	1.53	1.68	21.34	9.9
1.2	1.67	2.01	25.53	11.7
1.3	1.86	2.42	30.73	14.0
1.4	2.12	2.97	37.72	16.9
1.5	2.52	3.77	47.88	20.9
1.6	3.17	5.07	64.39	27.1
1.7	4.47	7.61	96.65	37.7
1.8	8.33	15.00	190.50	60.7

[a] An index of the increase in equilibrium global mean temperature associated with an effective doubling of atmospheric concentrations of carbon dioxide with 1 = 2.5°C.

[b] The expected marginal damage multiplier computed as instructed in equation (11b) according to the marginal cost structure defined by equation (7b) and the distribution of sea-level rise given by equations (8) and (9) for the doubling sensitivity specified in column (1).

[c] The complete marginal damage multiplier defined by equation (11a).

[d] Computed as the product of complete uncertainty multiplier and the middle Nordhaus (1991) marginal damage estimate of $12.70.

[e] Computed by comparing the marginal damage statistic computed in column (4) with the marginal cost of emissions reduction estimated by Nordhaus from published long-run energy analyses by regressing cost against the logarithm of (1 minus the percentage reduction). See footnote 10 and figure 5, Nordhaus (1991).

Recall that the multipliers of column (3) are translated into marginal damage estimates in column (4) by multiplying them by the middle value reported by Nordhaus (i.e., $12.70). A mean of $33.53 lies above the median of a distribution stretching from $5.84 on the low end to $190.37 on the high side. Column (5) finally relates these marginal damage statistics to the marginal cost of reducing carbon emissions, thereby suggesting a range of emission reductions which could prove to be efficient. The marginal cost figures used to support these reduction percentages are drawn, once again, from the original Nordhaus work. His figure 5, in particular, displays a composite marginal cost curve which is the result of a log-linear regression of emissions reductions against estimated cost run on data recorded in published long-run carbon emissions scenarios.[19] Emissions reductions supported by an efficiency criterion which equates their marginal cost with the expected marginal damage of allowed emissions run from a 3% reduction in cumulative emissions through 2050 (if a 1.5°C increase in the global mean temperature were associated with an effective doubling of carbon concentrations) to a 61% reduction in cumulative emissions (if doubling were to cause a 4.5°C increase). Figure 6 graphically displays the relationship between the $[T_d/T_0]$ index and efficient cumulative reductions in carbon emissions. The mean percentage reduction, roughly equal to 14%, lies slightly below the 15% reduction supported by the mean damage multiplier.

The specific numbers recorded in table 3 are certainly the product of the underlying structure. They are, however, quite insensitive to changes in the distribution of the doubling temperature which preserve the general gamma shape recorded in table 2 and dis-

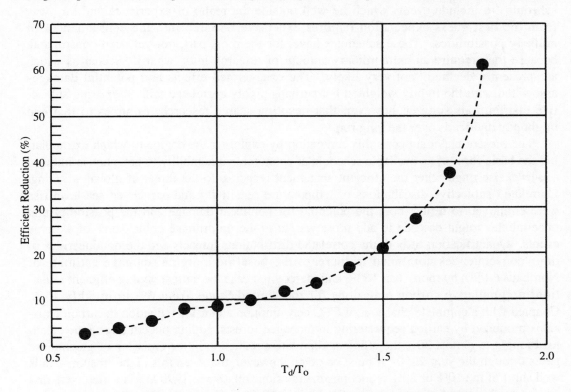

Figure 6. Efficient reduction targets for greenhouse gas emissions (percentage of unconstrained emissions through 2050 given specified doubling temperature changes). See table 3.

played in figure 5—i.e., fairly uniform density functions with some increased weight given to the lower half of the 1.5°C to 4.5°C range. Not surprisingly, however, the range of marginal damage estimates is extremely sensitive to the specific nonlinearity of the conditional damage function—the $\psi(SLR)$ function characterized in equation (7a). If that function were linear in sea-level rise, as an extreme example, then expected marginal damages would be a mere 23% higher than the Nordhaus estimate and support only a 7% emissions reduction.

CONCLUSIONS

Incorporating the subjective distributions of the uncertainty with which the future effects of global change phenomena are currently viewed is a difficult process, even before consideration of nonlinear impacts and dramatic surprises is added to the calculus. Proper evaluation across a range of possible futures requires, at the very least, some understanding of how a schedule of impacts and potential damage (net of adaptation but including the cost of adaptation) might be constructed over a range of foreseeable outcomes. Extending these schedules to include events which lie well outside the realm of experience and far away from the best-guess expectation requires expensive research into the consequences of unlikely possibilities. These schedules have, for the most part, not yet been constructed, because they require an extraordinary amount of research into "what if" scenarios which are presently deemed "not very likely." The exaggerated effects and potential damages associated with the lightly weighted but perhaps highly correlated tails of existing subjective distributions suggest, however, that devoting scarce research resources to that end might pay dividends over the long run.

The present paper supports this contention by exploring the degree to which extrapolating the basic form of an available schedule of economic vulnerability to greenhouse-induced sea-level rise might alter the efficient abatement response to the threat of global warming. Correlated subjective distributions of temperature sensitivity and associated sea-level rise were employed to reflect both the potential for nonlinear damage and the possibility that uncertainties might cascade to add more weight to the unfortunate coincidence of extreme events. Quantification of both the correlated distributions impacts and the nonlinearities of their consequences combined to increase the baseline marginal damage estimate of Nordhaus (1990) by more than 160% and thereby increase the corresponding efficient reduction in cumulative carbon emissions for the United States from 6% to roughly 15%. Coupled with a complete phaseout of CFC consumption and a 1% reduction in carbon emissions produced by carbon sequestering in managed forests, adding uncertainty to the calculus brought to almost 30% the efficient cumulative reduction in the emission of greenhouse gases through the year 2050. It must be noted, however, that even this higher response falls well short of the 10% or 20% reductions in emissions relative to 1990 levels which were discussed in the negotiations prior to the United Nations Conference on Environment and Development in June of 1992. Stabilizing emissions of carbon dioxide at 1990 levels would, for example, be consistent with a 50% reduction in cumulative emissions through 2050.

Besides adding some weight to the claim that more substantial emissions reductions should be supported by even the United States, the results suggest something more fundamental for the conduct of research into issues of global change. Analyses of these sorts of issues are typically so involved that careful attention can be paid to only a very limited number of possible futures. Best-guess scenarios have typically been selected, especially when time and resources reduce this number to one, but that might be a mistake. Taken qualitatively, the results reported above suggest that focusing on a scenario which describes something around the 75th percentile of potential economic damage might be a better choice—a better reflection of the potential significance of expected damage computed to include the coincidence of "bad news" tails.[20] At the very least, the construction of an uncertainty multiplier which measures the added expected cost of a wide range of extreme outcomes, is shown to be a productive tack with which to surround a "best guess" or "70th percentile" trajectory with some illustrative measure of the uncertainty with which the future is viewed.

It should be noted, of course, that all of the analysis presented here was built upon the house of cards which is oversimplification. The original Nordhaus paper abstracts dramatically from the complexity of the natural and social process which drive global change—figure 3 is a better portrait of its structure than the larger synthesis of figures 1 and 2. It also assumes that the composition of the United States economy in the year 2050 will look like the composition of 1981. It ignores investment possibilities, even as it looks at the tradeoff between current and future consumption, and it ignores other market failures which might increase or reduce the degree of efficient response to greenhouse warming. The extension presented above under "Quantifying the Uncertainty Multiplier" adds to these oversimplifications by extrapolating the shape of the aggregate damage function from the shape of the function relating sea-level rise to economic vulnerability—an expansion of the consistent composition assumption. It also linearizes some complicated structures and assumes risk neutrality in the social objective function.

Allowing more structural flexibility would certainly reduce damages, but adding risk aversion would certainly increase their welfare cost. The net effect of all of this simplification is likely to be significant, but the direction in which the uncertainty multipliers would move if more complete reflections of what might happen were included is unknown. The lesson that systematic inclusion of possible surprise events and nonlinear damages should make us more cautious in the protection of our health and well-being should not, therefore, be dismissed as the consequence of oversimplification.

ACKNOWLEDGMENTS

An earlier version of this paper was presented at the June 1992 meeting of the International Energy Workshop in Cambridge, Massachusetts, and was made available to Energy Modeling Forum 12. The research recorded here was sponsored by the Electric Power Research Institute and Connecticut Sea Grant (Number RLS/5). The author thanks

Marielle Yohe for her assistance in preparing the figures and John Weyant for a careful review of an earlier draft. All remaining errors, of course, reside with the author.

NOTES

1. These modest emissions reductions fall well short of fixing emissions at 1990 levels. Stable emissions along the baseline would, in fact, reduce cumulative emissions through 2050 by nearly 50%. Nordhaus (1993) contains revised estimates based on the next generation of his integrated model—DICE.

2. Figure 1 was prepared at a workshop hosted by the Aspen Global Change Institute in the summer of 1991; see Consortium for International Earth Science Information Network (1992). For figure 2, see Fisher (1988).

3. Baseline linearity across link A is not as rare as one might think. The IPCC Scientific Assessment (1990) shows, for example, a linear best-guess sea-level rise trajectory with a slope of 6 cm per decade surrounded by high and low scenarios characterized by slopes of 9 cm and 3 cm per decade, respectively. See chapter 9 of the assessment for more detailed descriptions. Linearity in link B could result from systematic response to predictable and anticipated impacts of predictable and anticipated climate change.

4. Impacts are given by $I = (1/60)^{1/2} t$, while damages are characterized by $D = (1/60)^{1/2} I = (t/60)$.

5. Estimates of economic vulnerability of the United States show a nonlinear correspondence with greenhouse-induced sea-level rise, primarily because inundation thresholds for economically valuable properties scattered around the coastline are crossed with increasing frequency as the seas rise even along a linear trajectory. See Yohe (1991a) for details.

6. Impact is now given by

    $$I = -(60)^{1/2} [1 + \exp\{(\ln(3)/60)t\}]$$

    which combines with schedule D of note 4 to produce

    $$D = 1 + \exp\{[\ln(3)/60]t\}.$$

7. Return now to schedule I from note 4, but let

    $$D' = -1 + \exp\{\ln(3)/60^{1/2}\}$$

    reflect damages so that they are, as a function of time, again given by

    $$D = -1 + \exp\{\ln(3)/60]t\}.$$

8. Actually, schedule III compounds I' from note 6 into D' from note 7 to reflect a nonlinear damage profile and a nonlinear impact correspondence. The result is:

    $$D = -1 + \exp\{-\ln(3) + [\ln(3)]\exp\{[\ln(3)/60]t\}\}.$$

9. Cline (1992) and Nordhaus (1993) represent more recent efforts.

10. The two sources cited by Nordhaus to support his empirical application are both reports to Congress, one issued in 1988 and the other issued in 1989. See U.S. Environmental Protection Agency (1988, 1989) for details.

11. The subsections entitled "The Nordhaus Model" and "Expected Marginal Damages of Emissions" are fairly technical and potentially tedious. The less interested reader can skip to "Quantifying the Uncertainty Multiplier."

12. See Nordhaus (1991) for a complete description of the model outlined briefly here and extended to incorporate uncertainty in the next section.

13. Nordhaus (1991 and 1993) employed, and Cline (1992) prefers, a logarithmic utility function with a relative risk aversion parameter of 1.

14. Setting the discount rate where intergenerational comparisons are involved is a highly contentious issue. See, e.g., Cline (1992) and Broome (1992) for strong opinions that β should be 0 in this case.

15. That is to say, two exponential schedules fit the vulnerability data well: one for anticipated greenhouse-induced sea-level rise under 80 cm and the other for anticipated levels more than 80 cm higher than otherwise.

16. See page 338 in National Research Council (1983), for example.

17. See Hoel, Port, and Stone (1971) for a general description of the gamma distribution. The density function for (α, λ) is given by

$$g(x) = x^{\alpha-1}e^{-\lambda x} \qquad \text{for } x > 0$$

$$g(x) = 0 \qquad\qquad \text{otherwise,}$$

with $\alpha > 0$ and $\lambda > 0$. The shape of $g(x)$ runs across its $x > 0$ domain from "relatively" exponential for $\alpha < 1$ to the mounded shape suggested in Figure 5 for $\alpha > 1$.

18. This value is computed, by definition, as the middle value reported by Nordhaus ($12.70) times the multiplier 0.46.

19. See footnote 10 in Nordhaus (1991) for a more complete description of this regression.

20. Yohe (1991b) suggests a method by which this sort of information about the distribution of future trajectories can be used to identify useful, "interesting" scenarios.

REFERENCES

Broome, John. 1992. *Counting the Cost of Global Warming* (Cambridge, UK: The White Horse Press).

Cline, William. 1992. *The Economics of Global Warming* (Washington, DC: Institute for International Economics).

Consortium for International Earth Science Information Network. 1992. *Pathways of Understanding: The Interactions of Humanity and Global Environmental Change* (Ann Arbor, MI: University of Michigan, University Center).

Fisher, Arthur. 1988. "One Model to Fit All." *Mosaic,* vol. 19, no. 3/4, pp. 52-59.

Hoel, P. G., S. C. Port, and C. J. Stone. 1971. *Introduction to Probability Theory* (Boston: Houghton Mifflin Company).

IPCC Scientific Assessment. 1990. *Climate Change—Final Report of Working Group I* (New York: Cambridge University Press).

National Research Council. 1983. *Changing Climate* (Washington, DC: National Academy Press).

Nordhaus, W. 1991. "To Slow or Not to Slow: The Economics of the Greenhouse Effect," *Economic Journal,* vol. 101.

_____. Forthcoming 1993. *Managing the Global Commons: The Economics of Climate Change* (Cambridge, MA: MIT Press).

Oerlemans, J. 1989. "A Projection of Future Sea Level," *Climate Change,* vol. 15, p. 151.

Shlyakhter, A., and D. Kammen. 1992a. "Estimating the Range of Uncertainty in Future Development from Trends in Physical Constants and Predictions of Global Change," Northeast Regional Center for Global Environmental Change, mimeo (Cambridge, MA: Harvard University).

_____. 1992b. "Sea-level Rise or Fall?" *Nature,* vol. 357, p. 25.

Wilson, R. 1988. "Measuring and Comparing Risk to Establish a *De Minimis* Risk Level," *Regulatory Toxicology and Pharmacology,* vol. 8, pp. 267-282.

U.S. Environmental Protection Agency (EPA). 1988. *The Potential Effects of Global Climate Change on the United States,* Draft Report to Congress (Washington, DC: U.S. EPA, October).

_____. 1989. *Policy Options for Stabilizing Global Climate,* Draft Report to Congress (Washington, DC: U.S. EPA, February).

Yohe, G. 1991a. "The Cost of Not Holding Back the Sea: Toward a National Sample of Economic Vulnerability," *Coastal Management,* vol. 18, pp. 403-431.

_____. 1991b. "Selecting 'Interesting' Scenarios with which to Analyze Policy Response to Potential Climate Change," *Climate Research,* vol. 1, pp. 169-177.

7
Assessing Climate Change Risks: Valuation of Effects

Anthony C. Fisher and W. Michael Hanemann

Other papers presented in this volume discuss and document the many and varied impacts of projected global warming—and the uncertainties about the importance of the impacts. We do not propose—we are not qualified—to contribute anything of substance to this discussion. Instead, it will be the purpose of this paper to speculate about some of the implications for economic valuation. Our emphasis will be on ways of thinking about and dealing with major impacts, even, or perhaps especially, where there remains disagreement about magnitude and timing. In doing so, we are not necessarily coming down on the side of those who foresee the gravest impacts. Rather, the question we wish to address is, what if they are right? What are the implications for valuation? For that matter, what if there is a reasonable chance they are right? How does this affect valuation?

The next section of the paper, "Damages from Global Warming: No Cause for Alarm," briefly reviews what is probably the best-known treatment by an economist of the damages to the United States from global warming—namely, the recent contribution by William Nordhaus (1991). The estimated damages, in fact, come from a report by U.S. Environmental Protection Agency (U.S. EPA, 1989); Nordhaus's chief contributions are estimation of the costs of slowing warming and development of a model balancing these costs against the damages. The section "Damages from Warming: A Critical Analysis," examines the appropriateness of the valuation exercise conducted by the EPA with regard to both which impacts were included and how these were analyzed. It seems fair to say that the EPA/Nordhaus view of the risks associated with global climate change is relatively sanguine, whereas others see higher probabilities of more drastic and even catastrophic impacts. The next section is about implications for valuation and policy choice if at least some elements of the less optimistic view are credible. Here we shall give particular attention to uncertainties, irreversibilities, and nonlinearities. Recommendations concerning the direction of future research are offered in the concluding section.

DAMAGES FROM GLOBAL WARMING: NO CAUSE FOR ALARM

The EPA estimates of damages from global warming associated with a doubling of carbon dioxide (CO_2), for the United States, are given in table 1 (taken from Nordhaus, 1991). The salient feature of these estimates is that they are very modest: just $6 billion, or 1/4% of gross national product in 1981 dollars. We shall discuss the individual categories shortly. But we first observe that it appears to be conventional, in the literature on climate change, to focus on the warming—variously estimated as between 1.5 degrees Celsius (°C) and 4.5°C—and related impacts associated with a doubling of CO_2. This custom may be due to the standard climate model projection of a <u>doubling</u> over the next several decades, and the understandable reluctance of climate modelers—or economists—to

Table 1. Estimated impacts of doubling of CO_2, United States

Sectors	Billions (1981 $)
Severely impacted sectors	
Farms	
Impact of greenhouse warming and	
and CO_2 fertilization	−10.6 to +9.7
Forestry, fisheries, other	Small
Moderately impacted sectors	
Construction	Negative
Water transportation	?
Energy and utilities	
Energy (electric, gas, oil)	
Electricity demand	−1.65
Non-electric space heating	1.16
Water and sanitary	−?
Real estate	
Land-rent component	
Estimate of damage from sea-level rise	
Loss of land	−1.5
Protection of sheltered areas	−0.9
Protection of open coasts	−2.8
Hotels, lodging, recreation	?
Total central estimate	
Billions, 1981 level of national income	−6.2
Percentage of national income	−0.26

Note: A positive number indicates a gain; a negative number indicates a loss.
Source: Nordhaus (1991, table 2.5).

make projections beyond a horizon of several decades. Or it may be due to model projections of an equilibrium at that level. We shall return to this issue, which may be the crucial one for our discussion—and, for that matter, for the wider policy debate—in the next section. In the table, two sectors, agriculture and forestry, are shown to have the potential to be severely impacted. No quantitative estimate of damages is given for the latter, only an indication that the value of the impact could be positive or negative—and is small in either case. With regard to agriculture, impacts could range (in round figures) from plus to minus $10 billion; the harmful effects of higher temperatures could be offset by the beneficial effects of higher CO_2 levels on plant and tree growth. Judging from Nordhaus's central estimate of the total damages, the net damage assigned to agriculture is about $0.45 billion (all figures in 1981 dollars).

Several sectors—construction, water transport, energy, and real estate—are classified by Nordhaus as having (in his opinion) the potential to be moderately impacted. Quantitative estimates are provided for only the last two, though it is indicated that the impact on construction will be positive. For energy, an increase in the need for space cooling is largely offset by a decrease in the need for space heating, so that the net cost is $0.49 billion. This leaves real estate, with a net cost of $5.29 billion, by far the largest component of the damages. These costs consist of the estimated value of land—about 4,000 square miles—lost to flooding and the estimated costs of protecting high-value property and open coastal areas.

Both the EPA and Nordhaus acknowledge that these estimates are incomplete, since they do not take into account impacts on ecosystems not immediately tied to the market economy or, for that matter, on nonmarket goods and services in general. Nordhaus cites one study as suggesting that global warming will provide major amenity benefits in everyday life (National Research Council, 1978); he concludes that "climate change will produce a combination of gains and losses with no strong presumption of substantial net economic damages" (Nordhaus, 1991, p. 46).

DAMAGES FROM WARMING: A CRITICAL ANALYSIS

Placing an economic value on the consequences of global climate change is an enormous challenge. It involves impacts not only on the supply of commodities that are traded in markets, but also on nonmarket goods. As noted above, these impacts could be quite profound, and they are likely to be widely spread over time and space. Quantifying these impacts in monetary terms surely pushes the existing techniques of economic valuation to their limits. Here, we want to review the types of analysis that were performed by the EPA's economists and its consultants in their report on the potential effects of global climate change on the United States and examine how successful they were in meeting this challenge.

Let us start by considering just what was valued. What impacts would you expect to find included in a study of global climate change? An answer is provided in the opening lines of the Executive Summary of the EPA report:

> Scientific theory suggests that the addition of greenhouse gases to the
> atmosphere will alter global climate, increasing temperatures and chang-

ing rainfall and other weather patterns . . . .   Such climate change could
have significant implications for mankind and the environment: it could
raise sea level, alter patterns of water availability, and affect agriculture
and global ecosystems (U.S. EPA, 1989, p. xxv).

Thus, four types of impact are to be expected: a rise in the sea level, a change in the
patterns of water availability, impacts on agriculture (including forestry), and impacts on
global ecosystems. In addition, there are the impacts on air quality and human health,
noted above. However, what was actually quantified in the EPA report was a somewhat
narrower set of impacts. This process of winnowing is described in these additional
excerpts from the Executive Summary:

After consulting with scientific experts, EPA developed scenarios for use
in effects analysis. Regional data from atmospheric models known as
General Circulation Models (GCMs) were used as a basis for climate
change scenarios . . . .The GCMs generally agree concerning global and
latitudinal increases in temperature, but they disagree and are less reliable
concerning other areas, such as regional changes in rainfall and soil moisture . . . .
Because the regional estimates of climate change by the GCMs vary considerably, the scenarios provide a range of possible changes in climate for
use in identifying the relative sensitivities of systems to higher temperatures and sea level rise . . . (Ibid., p. xxvii).

Thus, the EPA's quantitative analysis actually focused on the economic impact of the
rise in sea levels on real estate, the economic impact of higher temperatures on agriculture,
and the demand for electricity. Left out are the economic impacts associated with a change
in the patterns of water availability for agricultural, municipal, and industrial uses; impacts
on economic infrastructure in general; impacts on ecosystems; and impacts on human
health.

The Executive Summary notes some other limitations on the analysis. For example, it
notes that the scenarios assume no change in "the frequency of events, such as heat
waves, storms, hurricanes, and droughts in various regions, which would have affected the
results presented in this report" (Ibid., p. xxix). The scenarios also assume that climate
variability does not change from recent decades; this is because, when the results of the
GCMs were examined, "we found that no firm conclusions can be drawn about how global warming would affect variability" (Ibid., p. xxix). Another limitation noted is the failure to consider human adaptation to the effects of climate change through changes in population and technology.

All of this raises two sets of questions: (1) What are the reasons for these limitations?
Why did the EPA researchers exclude certain potential impacts, and what explains the
impacts that <u>were</u> included? (2) Do these exclusions matter—and do they matter enough

that the researchers should have done something to avoid them? With regard to the first question, obviously, part of the answer is that there were problems with the GCMs, including divergences with regard to their predictions of regional changes in precipitation, and limits on their ability to track intertemporal variability in climatic conditions on both a small temporal scale (say, within-year, seasonal variation) and a larger scale (say, year-to-year or decade-to-decade variation). But we do not believe that this was the only reason. We presume that another reason was the availability of economic models suitable for examining some impacts but not others. For example, we presume that this explains why the impacts of global climate change on agriculture and electricity demand were studied, while the impacts on the demand and supply of water for irrigation or municipal and industrial uses were not.

Several decades ago there was a popular song with these lines: "If you can't be with the one you love, then love the one you're with." For researchers, the equivalent of this advice to the lovelorn might be something like: "If you can't measure what's important, then play up the importance of what you can measure." There are at least some grounds for concern that this is happening here. Before considering them further, however, we want to examine how the EPA researchers treated the impacts that they did quantify, because we have reservations about some aspects of the methodologies that were employed.

The Cost of Sea-Level Rise

We will start with the impact of sea-level rise on real estate. As noted earlier, part of the economic impact is measured by the costs of protecting open coasts and certain sheltered areas from flooding, while the other part is measured by the value of land that is not saved from flooding. Here, we focus on the latter. The methodology is illustrated in figure 1, which is taken directly from Yohe (1989). Figure 1 shows a hypothetical property-value gradient for land as a function of its proximity to the shore. For concreteness, suppose that there are seven tracts of land running from tract A, which is immediately adjacent to the shore, to tract G, which is furthest inland. The property values of these tracts decline with distance from the shore: they are highest ($100,000 per lot) for seafront land in tract A, next highest (at $90,000 per lot) in tract B, and eventually they stabilize at $50,000 per lot in tract E and other inland tracts (F, G).[1] Now, suppose that tract A is flooded because of the rise in sea level. There are two components to the resulting estimate of damages: the value of the lost structures that had formerly stood on tract A, and the value of the land that is now lost to human uses. Yohe's analysis of the latter runs as follows:

> Were the sea to rise so that the first lot were lost, then the second lot would become a shoreline lot and assume the $100,000 value originally attributable to the first. The value of the third lot would climb to $90,000, and so on. The community would, in effect, lose the economic value of an interior lot located initially more than 500 feet from the shoreline. The true

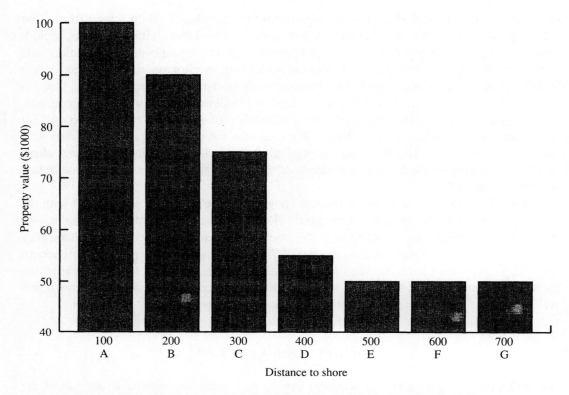

Figure 1: The property value gradient for land adjacent to the shoreline [Source: Yohe (1989)].

economic loss would be the equivalent of a $50,000 lot instead of the shoreline $100,000 lot; there would be a distributional effect, to be sure, but the net social loss would be $50,000 (Ibid., p. 4-4).[2]

For small marginal losses of shoreline land, this is clearly a valid approach—with the caveat that it depends on the use to which the new shoreline will be put. A sandy beach, for example, could take centuries to form. With larger nonmarginal changes, two additional problems could arise. First, the analysis assumes that there is no shortage of land on which the owners of the flooded tracts can relocate—in effect, that there is an infinitely elastic supply of inland tracts like E, F, and G. This is a reasonable assumption on a featureless plain like, say, Kansas. It is more problematic if there are natural barriers—mountains, lakes, or rivers—which limit the space available for human occupation, as might happen along the coast of California, for example.[3] In that case, when some land is inundated, the remaining land becomes more scarce, its value increases—even for inland tracts like E, F, and G—and the damage is understated if one prices the lost land at the preflood value of interior tracts.

Second, a nonmarginal loss of coastal land presumably involves the destruction of a significant amount of public infrastructure—roads, bridges, seawalls, etc.—which then has to be rebuilt along the new shoreline. It is not clear that the costs of replacing this infrastructure have been captured in the EPA/Nordhaus estimate of lost land values. Two separate questions are being raised here: (1) To what extent do the market values of private property incorporate the value of adjacent public infrastructure that is a complement to the private property? (2) To what extent do property values reflect the replacement cost of infrastructure capital? With regard to the first of these questions, the answer would seem to depend on the nature of the public infrastructure. Access to a road might be expected to increase the value of nearby private land; therefore, the value of the road should be captured, at least partly, in the value of the private land. However, moving costs or other impediments to free mobility might prevent the complete capitalization of the value of public infrastructure in private property values. Moreover, the road might be perceived as detrimental—due to noise, say, or air pollution—and, therefore, it might reduce the value of adjacent private land. The situation is similar with a sewage treatment plant. The point is, what is capitalized is the value of the public infrastructure to residents, not necessarily its cost.[4] With regard to the second question, there is some evidence that the replacement cost of public infrastructure has risen over time relative to other types of capital and relative to the prices of other goods and services, because it involves relatively labor-intensive construction activity and because it has not experienced the same rate of technical progress as the supply of other goods and services. Therefore, the gap between historic and replacement cost may be especially large.

There is a larger question here about how a community might adjust to the loss of physical capital when sea level rises. The analysis described above implicitly assumes a smooth and efficient adjustment process that minimizes the damage: after tract A in figure 1 is flooded, tract B becomes shoreline, and everybody dutifully moves over one tract to the right. Regrettably, this does not describe the world as we know it. For example, the failure of state and local governments in the United States to regulate floodplain development is notorious. Another example is somewhat closer to home: there have been about a half dozen wildland fires in the Berkeley-Oakland hills since 1930, culminating in the tragic fire of October 1991. Each time, residents rebuilt after the fire in the same location, and few lasting efforts were undertaken to eliminate the future threat of a fire or to mitigate its potential impact. In one prosperous community south of San Francisco, a ban on shakes and other flammable roofing materials was imposed after a serious fire in 1981, only to be repealed three years later due to public pressure. Such bungled responses raise the costs of natural disasters. The inability to respond effectively is a cost of doing business that has to be included in any estimate of the economic impacts of global climate change.[5] We are not suggesting that the government ought to prohibit all risky activities—though there may be a case for prohibition, or other control, where externalities (wood shake roofs cause a fire to spread) are present. Rather, the point is that where many people are involved in a decision, such as reconstruction in a region following a natural disaster, the transaction costs are likely to be large and need to be taken into account in assessments of the damages. For all of these reasons, we feel that there are grounds for questioning the EPA/Nordhaus estimates of the impact of sea-level rise on real estate. We turn next to their estimate of the impact of a doubling of CO_2 on agriculture.

The Costs to Agriculture

The EPA report actually presents four estimates of these economic impacts, which are reproduced in table 2. Two of the estimates incorporate the impacts on crop yields not only of higher temperatures but also of higher CO_2 levels, the latter involving beneficial effects associated with increased photosynthesis and improved water-use efficiency. These estimates, which Nordhaus employs for his own analysis, are either a net loss of consumer's surplus plus producer's surplus amounting, on an annual basis, to $9.7 billion (in 1982 dollars), or a net gain amounting to $10.6 billion.[6] When the effects of higher CO_2 levels are omitted and one focuses on the consequences of higher temperatures alone, the EPA estimates a net loss ranging from $5.9 billion to $33.6 billion.

Table 2. Aggregate economic impacts on agriculture ($ billion/year in 1982 dollars)		
	No CO_2 effect	With CO_2 effect
Goddard Institute GCM	-5.9	+10.6
Geophysical Fluid Dynamics Lab GCM	-33.6	-9.7

Note: GCM = General Circulation Model.
Source: U.S. EPA (1989), table 6-4, p. 104.

The other factor that accounts for the different estimates in table 2 is a divergence between the predictions of the two GCMs that EPA used regarding regional climate changes; in each case, the higher damage estimate is associated with the temperature scenario generated by the Geophysical Fluid Dynamics Laboratory GCM, while the lower damage estimate comes from the temperature scenario of the Goddard Institute for Space Studies GCM. These models differ with regard to what they predict about the reduction in rainfall in the Southeast (which lowers crop yields, especially for nonirrigated crops) and the increase in temperature in the very northern areas such as Minnesota (which extends the frost-free growing season).

The differences between the two GCMs are actually greater than the differences between the with-CO_2 and without-CO_2 scenarios. Obviously, we have no competence to discuss the former. With regard to the latter, we do wonder about the magnitude of the beneficial effects attributed to higher CO_2 levels. Indeed, the EPA report itself warns that they may be exaggerated:

The direct effects of CO2 in the crop modeling results may be overestimated for two reasons. First, experimental results from controlled environments may show more positive effects of CO_2 than would actually occur in variable, windy, and pest-infested (weeds, insects, and diseases) field conditions. Second, since the study assumed higher CO_2 levels (660 ppm) in 2060 than will occur if current emission trends continue (555 ppm), the simulated beneficial effects of CO_2 may be greater than what will actually occur (Ibid., p. 100).

Suppose the increase in CO_2 does lead to increased yields. What factors might counter this beneficial effect? One possibility is that the same effect will be seen for weeds and pests (Daily and coauthors, 1991). Further, the related warming will widen the range of pest species, in particular by increasing overwintering survival (Gleick, 1991). A dramatic illustration of the potential for this effect was given just last year by the sudden invasion of California's Imperial Valley, a prime agricultural area in the southern part of the state along the Mexican border, by the apparently omnivorous silverleaf whitefly. A halt to the tiny pest's devastating march through the Valley was achieved only after some cooling in surface temperatures. And it is only the modestly cooler temperatures now prevailing in California's great Central Valley that offer protection to the $19 billion agriculture industry there. Warming would also widen the range of crop and livestock disease organisms—in particular, very debilitating ones now limited to tropical regions (Gleick, 1991; U.S. EPA, 1989, p. 94).

So far, we have considered crop yield considerations that could make the damage estimate of $9.7 billion significantly too low. What about the rest of the economic analysis performed by EPA's consultants for the agricultural sector? Are there any other factors that could influence the damage estimates one way or the other? Three such factors come to mind. The first concerns the specification of the substitute uses of land that enter into the calculation of damages from climate change. It is inevitable that the damage estimates are sensitive to what substitute uses are included in these calculations. Errors can be caused both by excluding substitutes that should be included and by including substitutes that should properly be excluded. As Mendelsohn, Nordhaus, and Shaw (1992) have observed, to the extent that the EPA's analysis focused on measuring the reduction in yields for specific crops, such as wheat or corn—i.e., ignored the possibility that farmers might switch to other crops or other uses of land that are not as greatly affected by climate change—its analysis will overestimate the damages from climate change. Conversely, however, if one includes in the analysis alternative crops or alternative uses of land that are not, in fact, part of economic agents' choice sets, the result would be to underestimate the damages. This is a problem, for example, with programming models of agricultural production in California and some of the other western states which tend to include an extremely large array of crops in the choice set—larger than is likely to apply to any individual farmer or group of farmers. The point is that ensuring the correctness of the specification of the choice set requires effort and is something that researchers frequently tend to overlook, but if it is not done it can impart a significant bias to damage estimates.

Another factor is the time dimension of the analysis. Clearly, this presents an enormous challenge: the EPA's researchers were using models calibrated to 1980-83 conditions to predict economic impacts in agricultural markets in 2030-60, some 50 to 80 years later. These are static models, with no change on the supply side or the demand side. The EPA report notes that some of the changes that might be anticipated on the supply side could mitigate the economic effects of global climate change. It points out that changes such as higher yielding crop varieties, chemicals, fertilizers, and mechanical power have historically enabled the agricultural sector to boost yields, and it estimates that, if the same rate of yield increase experienced from 1955 to 1987 were to continue into the future, most of the adverse impacts of climate change could be offset. On the other hand, it points out that changes on the demand side—increasing food demand from higher U.S. and world populations—could aggravate the economic loss from climate change. These are certainly valid points, and we have nothing to add to them.

However, there is another feature of the model used for the agricultural impact analysis that we believe could affect the cost estimate, but that has received relatively little attention. In effect, the model assumes malleable capital that can be shifted costlessly in response to perfectly anticipated economic changes. It is a static, spatial-equilibrium model; as the EPA report notes, "it simulates an equilibrium response to climate change, rather than a path of future changes" (Ibid., p. 103). Thus, it doesn't say how long the economy takes to move from one equilibrium to another, nor what the economic costs are during the out-of-equilibrium phase. It is our understanding that the model does not include adjustment costs, nor indeed capital costs generally—just the variable costs of production. There is an implicit assumption that the unmeasured disequilibrium costs are small in magnitude relative to the measured costs of switching from one long-run equilibrium to another.

Indeed, this assumption underlies most of the EPA/Nordhaus treatment of the impacts of climate change: their analysis focuses on changes in the annual flow of goods and services, rather than on changes in the stocks of economically significant capital—physical, human, or natural. We believe that this is an important omission. It is our hunch that some of the most important impacts of climate change arise because of the effect on capital stocks which, if not destroyed, are rendered prematurely obsolete. The costs of these effects depends critically on their timing relative to the normal replacement cycle of the affected capital. If the capital was going to be replaced anyway and the effects of climate change are well anticipated, an adjustment to climate change can be incorporated with minimum cost and disruption. If the capital was not due for replacement—or is difficult or costly to replace—then the costs are much greater. In this regard, it is quite possible that the costs of obsolescence prematurely imposed on human or natural capital could be significantly larger than for physical capital. Take, for example, the effects of rising temperature and related declines in soil moisture in areas currently well suited to grow grains and other crops, such as the American Midwest and Central Europe (Cline, 1991; Daily and coauthors, 1991). These impacts could require a drastic and slow adjustment given the existing investment in physical infrastructure, extension and credit institutions, etc. Moreover, it is unlikely that the natural capital stock—the soil that has evolved a set of characteristics suited to these crops—could be similarly "adjusted" in any time frame of interest to human beings. Productivity characteristic of the American grain belt may not be transferable to the thin acidic soils of the Canadian shield (Brown and Young, 1990).

We emphasize the distinction between this "capital-oriented" view of the impacts of climate change and the conventional "flow-oriented" view because we believe that the social value of a dollar of added investment cost may exceed the social cost of a dollar of reduced flow of goods and services, for at least two reasons. First, there is the potential for a "crowding out" of conventional and productive investment in order to make way for the replacement investment induced by climate change. If there are constraints on the supply of savings, this could have a long-run cost in terms of reduced economic growth. Second, to the extent that climate change requires a <u>collective</u> response—for example, a collective decision to relocate in upland areas—there is some likelihood that imperfect coordination will inflate the costs of adjustment, as in examples of floodplain zoning and rebuilding after the Bay Area fires cited above.

This distinction becomes even more significant when one considers the possibility, discussed below, that the change in climate will not stop at a doubling of CO_2 but could involve even larger increases. The greater the disruption of physical and natural systems, the greater the economic impact in terms of premature obsolescence of valuable physical, human, or natural capital, and the greater the potential downward bias from employing a flow-oriented approach to measuring damages.

Unmeasured Impacts

We have spent some time discussing how the EPA researchers measured some of the costs of climate change that they did quantify. What about the costs that were not quantified, and how do these relate to the capital-oriented approach to damage measurement that we have advocated? Of the various impacts that are listed but not quantified in the EPA report, two stand out as especially relevant here—the impacts on water availability and urban infrastructure. We want to emphasize the importance of the former from the perspective of the western United States. To someone living on the East Coast, changes in the timing or regional incidence of precipitation may seem of secondary importance. In the arid West, however, they are critical. In California, for example, on a statewide and annual basis there is more than adequate rainfall at least for the current population. However, two-thirds of the precipitation falls north of Sacramento, while two-thirds or more of the population have always lived south of Sacramento; similarly with the timing of precipitation, which occurs almost entirely in the winter, while peak demands for agricultural, in-stream, and even some urban uses occur in the late spring and summer.

Preliminary studies suggest that these imbalances will be exacerbated by climate change, which is expected to result in an increase in winter rainfall and a reduction in the snowpack, leading to less runoff in the late spring, when irrigation needs are highest. The solution to the imbalances has always been to store water in aquifers or in above-ground reservoirs for carryover to the summer months and for transport to the areas of use. But this traditional approach to water resource management is now under severe challenge from several sources, including fiscal and legal. Ever since the Carter administration, there has been a marked decline in the willingness or ability of the federal government to subsidize the construction of new water projects; at the same time, their costs have escalated dramatically. Legally, there has been a substantial shift, starting with the Mono Lake decision in 1983 upholding the use of the Public Trust doctrine to disrupt otherwise established water rights to divert water for off-stream uses. This was reaffirmed and expanded in the 1986 decision in *United States v. State Water Resource Control Board,* which set aside the board's 1978 decision on water diversions from San Francisco Bay/Delta on the grounds that it gave insufficient attention to in-stream needs and was not based on a balancing of all needs within the basin, in-stream as well as off-stream. Accordingly, while we do not regard the problems posed by climate change for water supply in the West as insuperable, we do anticipate that the costs of overcoming them could be very substantial.

Other impacts that were left unquantified in the EPA study, such as the destruction of natural ecosystems, are obviously likely to have a substantial cost. Wetlands and coral reefs are particularly at risk. Coastal wetlands, blocked by dikes, roads, and other impediments, may be unable to migrate inland to escape rising sea level, and coral reefs are exceptionally sensitive to changes in water temperature. Losses to either or both of these ecosystems would adversely affect productivity of ocean fisheries, which depend on them for nursery grounds and food supplies (Daily and coauthors, 1991). Terrestrial ecosystems would also be affected. From just a doubling of CO_2, the southern boundary of forest ranges could move northward by 700 kilometers (km). Since the known historic rate of migration is just 50 km per century, very substantial loss of forest is indicated, with an associated increase in species extinctions (U.S. EPA, cited by Cline, 1991). Ecologists are in general agreement that these and other warming-related changes would accelerate an already worrisome loss of the biodiversity that plays a crucial role in sustaining agricultural productivity (through continuing infusion of wild strains), the pharmaceutical industry, and, most important, a wide range of life-support systems, ranging from cycling of nutrients to disposal of wastes (Harte, 1991; Ehrlich and Ehrlich, 1981).

Perhaps less grand, but more obvious, are potential impacts on human health and well-being. Noteworthy here are an increase in the frequency and intensity of heat waves experienced in temperate regions such as most of the United States, and the spread of diseases now confined to the tropics (Harte, 1991; Daily and coauthors, 1991). Very small increases in yearly mean temperatures could permit the extension of tropical parasitic diseases into Europe and North America (Haines, 1990). Moreover, the EPA study indicates that there could be adverse impacts on human health due to hotter temperatures, greater variability in temperature, and increased air pollution resulting from climate change.[7]

There is a further point about the unquantified impacts on air quality and human health. As Ayres and Walter (1991) have observed, the human actions that give rise to global warming—most important, the combustion of fossil fuels—also have significant negative impacts on air quality and human health. While those impacts are not part of the costs of climate change, the benefits from reducing those impacts certainly are joint products of actions taken to avert or mitigate climate change. Looked at this way, the damages to air quality and human health resulting from the combustion of fossil fuels are relevant information in any assessment of policies for averting or mitigating global climate change. Ayres and Walter have suggested that, on a per-ton-of-CO_2-equivalent basis, damages could be an order of magnitude larger than the costs quantified in the EPA/Nordhaus analysis.[8]

We could continue with these examples of impacts from global warming associated with a doubling of CO_2. But just from what has already been presented, it seems plausible (at least) that damages may be more widespread, and more severe, than indicated in the EPA estimates. More important, the really crucial issue, as Cline (1991) has emphasized, is whether it is appropriate to limit our focus to the impacts associated with a doubling of CO2, as opposed to some larger increase.

Beyond a Doubling

Is there reason to believe that a doubling would represent an equilibrium? Not that we can discover. Perhaps the most authoritative recent report, that of the Intergovernmental Panel on Climate Change (IPCC, 1990), projects a doubling by the year 2025, with an associated warming of 2.5°C, assuming "business as usual." In most accounts, the story stops there. But the IPCC goes on to project a warming commitment of 5.7°C by the year 2100. The scientific literature on what happens after that is relatively sparse. Cline projects carbon emissions out to the year 2275 on the basis of assumptions of low growth rates (claimed, though not presented) for population and economic activity, and known availability of fossil fuel resources at reasonable recovery costs. The projected carbon emissions imply an increase in atmospheric concentration of CO_2 to almost eight times preindustrial levels. These emissions are augmented on the basis of IPCC estimates of the relationship between carbon and other greenhouse gas emissions. The result (Cline, 1991, p. 914):

> Over a horizon of 250-300 years the stakes of global warming are closer to
> a central estimate of 10°C rather than the 2.5°C associated with the bench-
> mark doubling of CO_2 which has so far dominated both scientific and poli-
> cy discussions.

One may, in fact, wonder if even this is a sufficiently distant horizon; has an equilibrium been reached, or is the process explosive? Cline suggests, citing a study by Sundquist (1990), that on time scales of 250 to 300 years, mixing into the deep ocean becomes important, opening a much greater sink. Of course, this is not certain; the process may be explosive. Or the indicated concentrations of CO_2 and other greenhouse gases may not be reached; the ocean sink may open sooner, enhancement of cloud albedo by sulfur dioxide air pollution may interfere (Booth, 1990), and so on. In our judgment, a scenario for a much greater level of warming (than that associated with a doubling of CO_2) has been constructed that has at least some credibility, though on a time scale beyond that with which economists (and perhaps climate modelers) are usually comfortable.

The usual objection would be that projections beyond a few years, and certainly beyond a few decades, are unneeded and unwarranted, given discounting and given the uncertainties involved (see, for example, Beckerman, 1991, who also argues that the near-term effects are not likely to be significant). We do not share this view. The impacts associated with a 10°C warming are likely to be very much greater even than those discussed earlier in this section. If they are, discounting will not make them go away. Alternatively, some might argue that discounting catastrophic impacts on future generations is simply immoral. Either way, it seems to us that discounting cannot be relied on to free us from the obligation to estimate, as best we can, impacts beyond the usual time horizons of economic models. With respect to uncertainties, the implication for valuation is not that we should throw out damage estimates to which they are attached. Instead, as we shall suggest in the next section, the estimates may, more appropriately, be augmented by option values and risk premiums.

We stated just above that the impacts associated with a 10°C warming are likely to be very much greater than those associated with a 2.5°C warming. What, specifically, might happen? Here we are on very thin ice, since most of the literature focuses on impacts of the lesser warming. Cline (1991) sketches some possibilities. Looking first at agriculture, CO_2 fertilization effects are less than linear and contribute little beyond the first doubling (U.S. Department of Agriculture, 1989). Moreover, yields collapse at temperatures in the range of 35° (fine grains) to 45°C (coarse grains), temperatures which would be routinely reached in the American grain belt. (Recall that even a seemingly modest increase in global mean temperature implies an increase in the frequency and severity of local or regional heat waves.) Thus—shifting the focus from agriculture—Cline calculates that the number of major U.S. cities experiencing average daily maximum temperatures in July of 38°C or more would rise from 2 to 44. We have already noted that the more modest warming associated with a doubling of CO_2 could have a serious impact on human health. Presumably, the impact would be multiplied by a 10°C warming with its frequent and prolonged episodes of extreme high temperatures.

The increase in sea level associated with just a doubling of CO_2 is generally estimated in the range of 30 to 60 centimeters (cm) (Nordhaus, 1991; Daily and coauthors, 1991). EPA's real estate damage estimates (their only substantial damages; see table 1) are based on a 50-cm rise. The IPCC projects a rise of from 31 to 110 cm, with a central estimate of 66 cm, by the year 2100, but this projection assumes that the Antarctic is a sink for water, rather than a source (melting ice around the edges is more than compensated by increased snowfall in the interior). Cline suggests that at the temperatures expected to prevail, the Antarctic, with 90% of the world's ice, will instead become a major source. By the year 2100 it could contribute 220 cm to a total sea-level rise of 367 cm (Hoffman, Wells, and Titus, 1986). The estimates by Hoffman and coauthors are for the high end of a range, and their range lies above the IPCC's own, but the question is not so much who is right about the medium term, but what happens in the longer term. Climate models suggest that relatively greater warming takes place at the higher latitudes; for an average warming of 10°C, regions around the poles could be expected to experience a warming of 15°C to 20°C. At these temperatures, it is at least possible that enough ice would melt to establish the Antarctic as a net source, and probably a major one, as in the projections by Hoffman and coauthors. Cline concludes that an increase in global mean temperature of 10°C would result in a sea-level rise of at least 400 cm, or 4 meters. A rise of 1 meter would eliminate 3% of the earth's land area and a larger percentage of its cropland (Rosenberg and coauthors, 1989), including over 30% of the most productive cropland (Wilson, 1989). To our knowledge, there are no similar estimates of the impact of a 4-meter rise, but it seems safe to say that it would be catastrophic. It is worth noting that this is a compelling example of a nonlinearity in the underlying physical process that has implications for economic analysis, as we shall spell out in the next section.

There is one other aspect of the predicted impacts, whether associated with a doubling of CO_2 or an 8-fold increase, or something in between, that deserves mention before the discussion of economic implications. The impacts are essentially irreversible on a time scale of interest to human beings. Perhaps this is too strong. Concentrations of CO_2 and other greenhouse gases have a residence time in the atmosphere that is measured in decades, even centuries. Making the extreme and unrealistic assumption that emissions are totally eliminated, concentrations would approach preindustrial levels only after several decades at the earliest. And even if climate change were reversible in the short term, important impacts of the elevated concentrations, such as accelerated loss of species or inundation of coastal improvements (roads, rails, buildings, power plants, etc.) would be irreversible.

THEORETICAL IMPLICATIONS FOR VALUATION AND POLICY

There are three considerations that seem relevant to the valuation of climate change risks that follow from our discussion to this point. One we have just noted is that certain kinds of decisions or actions are, for all practical purposes, irreversible. Another is that some of the potential impacts are catastrophic. The two may be related—the prospect that a loss or cost will be endured in perpetuity may tip an otherwise modest impact into catastrophe. What sort of actions are irreversible? We have just seen that emissions of CO_2 (and chlorofluorocarbons) qualify, by virtue of their very long residence times in the atmosphere. Burning a ton of coal today contributes to an irreversible warming commitment for the future. Similarly, cutting and burning an acre of rainforest (without replanting) contributes to the warming commitment. Investment in certain kinds of fixed facilities also contributes, in a somewhat different way. Siting a road or a power plant along a coast subject to flooding in the event of sea-level rise is "storing up" future damages, with some nonzero probability.

A third important characteristic of the value of damages that would follow from actions that lead to warming commitments, or even the siting of potentially affected facilities, is that it is uncertain. We may know that a doubling of CO_2 implies a warming commitment of 2.5°C and even that this in turn will result in certain kinds of damage, as for example, to agricultural productivity in a region. But how much? And even if we are certain about the physical impactsæeven if we know, for example, that the production of wheat will be reduced by 25%, what is this worth? An answer to this question clearly involves knowledge of demand, or preferences, for future goods. Information relevant to an answer presumably improves as the "future" gets closer.

Elsewhere, we have shown that when an action or decision has the characteristics that it is irreversible, that future costs and benefits are uncertain, and that information about the costs and benefits improves with the passage of time, there is a value ("option value") to refraining from the action during the current period. Alternatively, there is a cost, in the shape of a reduction in the ability of the decision maker to realize the value of information, attached to going ahead with the action in the current period (Fisher and Hanemann, 1986a, b, 1987). In one of these studies (1986a), we tried to calculate the option value of preserving a site that was later found to contain a potentially useful plant species, a wild relative of corn that is also a perennial. The calculation was done on the basis of some empirical information (about demand and supply functions for corn) coupled with several assumptions (about probabilities of successful hybridization and of realized values of alternative uses of the site), and so does not qualify as a true empirical application. Moreover, even if it did, it is about the benefits of preserving just one species, in just one site. With these caveats, we think it is worth noting that the calculated option value turned out to be substantial in relation to the conventionally estimated expected benefits in the example: around one-third of the expected benefits of preserving the site, and from one-tenth to two-thirds of the hypothetical expected benefits of developing, depending on what was assumed about these benefits. Clearly, we are a long way from knowing how to attach costs representing the foregone value of information to all of the actions leading to global warming commitments. But we should at least be aware that we may be leaving out a substantial part of the cost of such actions.

Now, let us consider the implications of catastrophic impacts. Here, we would like to propose a very simple framework that is best explained with the aid of a couple of diagrams. Figure 2 shows total damages from global warming, on the vertical axis, plotted against a measure of warming on the horizontal axis. (The horizontal axis could alternatively represent emissions, or concentrations, of greenhouse gases.) Of course, all of this is hypothetical, as we have not done any measuring or estimating. Nevertheless, we probably know enough about the phenomenon of environmental disruption generally to specify some properties of the functional relationship in this case. Thus the curve slopes up and to the right, indicating that damages increase with warming. Further, the curve is convex from the origin to point A, and again beyond A. Ignoring for the moment what happens at A, the convex functional relationship between warming and damages indicates that damages not only increase with warming, but do so at an increasing rate. Convexity of the damage function is a fairly standard assumption in environmental economics, presumably based on some evidence, as well as intuition. Yet the curve in figure 2 taken as a whole—

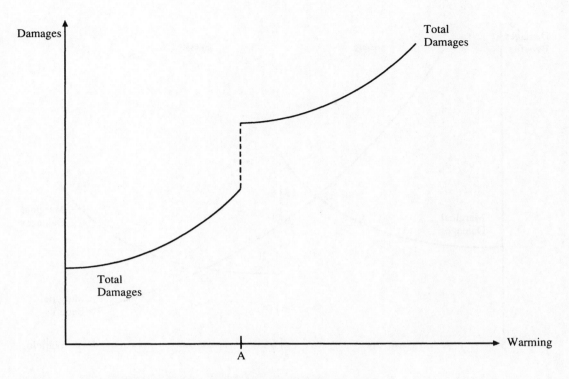

Figure 2: The behavior of damages from global warming.

that is, including the sharp increase or jump at A—is nonconvex. What is different here is clearly the existence of a jump in the damage function—this is the way we represent a catastrophic impact. There may well be more than one such jump in reality, though one is sufficient to illustrate the argument here.

Figure 3 translates total damages to marginal. Marginal damages increase to A, fall at A, and then begin to increase again. The significance of the resulting break in the marginal damage function is clear when it is displayed along with the assumed marginal benefit function on the figure. The benefit function slopes down and to the right, indicating that the benefit of increasing the level of activities that lead to warming is diminishing. Another way of understanding the behavior of this curve is to read it from right to left, in which case it represents the (rising) cost of controlling emissions, reducing concentrations, or mitigating impacts. Notice that the curves intersect three times around A: at the points labeled E and F, and again at point G. More generally, whenever the damage function becomes highly nonlinear or discontinuous, as at A, the benefit function may intersect it more than once, as here. The intersection at F has no welfare significance, since net benefit can be increased by moving to either E or G. Strictly, a choice between these points would require a benefit-cost analysis to determine whether the losses represented by area I in the figure exceed the gains represented by area II. If area I is larger—and perhaps also

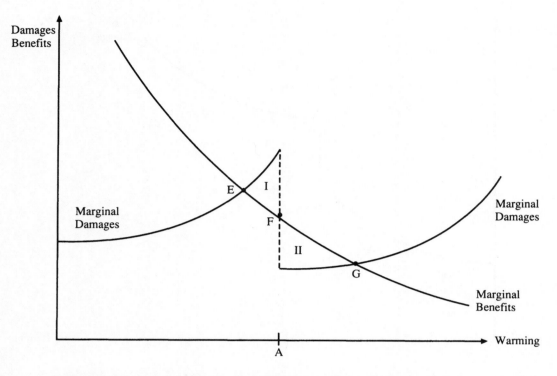

Figure 3: The behavior of marginal damages and benefits from global warming.

in the likely case that sufficiently precise determination is not possible—the implication for policy is that emissions of greenhouse gases should be controlled to some point before A, to avoid a catastrophic warming commitment with a margin of safety. This point is strengthened by recognition of the uncertainty surrounding estimates of damages from warming, in particular of the uncertainty about where the damage function becomes highly nonlinear or discontinuous, and the prospect that better information will make possible better estimates with the passage of time. An option value then attaches to refraining from actions that increase the warming commitment to a point near A.

In this context, we want to express our concern over the disconnect that is apparent in much recent discussion between the valuation exercise and the consideration of specific policies for mitigating climate change. The implicit assumption is that the damage estimation and the policy analysis can be conducted on two separate tracks: i.e., that damages can be assessed without knowing what control policies or, indeed, what levels of control will be selected after the damage figures have been developed. If economic analysis were a free good and if one could readily obtain a perfect and comprehensive analysis of every impact, this would be a harmless approach. But this isn't so, and it becomes necessary to focus the damage assessment and quantification on the control actions and control levels that are relevant for the policy debate. The discussion tends to focus implicitly on a particular policy tool—viz., the imposition of a uniform carbon tax on CO_2-emitting activities. For the purpose of that policy, all one needs to know is the marginal damage of an additional unit of CO_2 equivalent. A carbon tax may well be desirable, but it is not necessarily the only policy action that needs to be taken to deal with climate change, or even the best one. The literature on the relative merits of standards and taxes, or marketable permits and taxes, is relevant here. In particular, where damage functions exhibit the kinds of sharp nonlinearities and discontinuities sketched above, we know, following the work of Weitzman (1974) and others, that (other things being equal) quantity controls will tend to be preferable to charges. Thus a system of standards and marketable permits will be superior to a tax. Of course much more work needs to be done to determine the optimal choice or mix of policy instruments to slow global warming. Our point is simply that there is a link between valuation and decision, and that these exercises ought not be conducted entirely on separate tracks.

CONCLUSIONS: DIRECTIONS FOR FUTURE RESEARCH

We do not put ourselves forward as natural scientists, qualified to pronounce on the questions of the physical impacts of global climate change. We have presented some observations and conjectures, taken from the scientific literature, that in our judgment credibly suggest the possibility of much larger impacts than those with which most economists are likely to be familiar. These impacts, in turn, can be expected to result in "badly behaved" damage functions: nonconvexities in total damages, severe nonlinearities or discontinuities in marginal damages. We have suggested several reasons why the marginal damages may not be constant, independent of the level of climate change. First, the physical and biological damages—for example, the damages to wetlands or aquatic ecosystems—may well be a nonlinear function of carbon emissions or the rise in sea level. Second, the economic costs of adjustment are likely to be a nonlinear function of the magnitude of the adjustment, although here the nonlinearity could perhaps go in either direction. On the one hand, it might be argued that the greater the physical change, the greater the likelihood that people will recognize it and factor it into their future planning. On the other hand, the greater the change, the greater the need for collective action to deal with it, and the greater the difficulties and costs of coordinating this response. Moreover, the greater the change, the greater the potential "crowding-out" consequences for other needs as a result of the requisite infrastructure investment to deal with climatic impacts.

We conclude with two observations. First, since knowledge of the damage functions—in particular of the regions where they are badly behaved—is crucial for policy regarding emissions of greenhouse gases, we would urge support for research that focuses on what happens beyond the doubling usually assumed for concentrations of CO_2. Second, we would urge that more of the economic research be focused on the potentially very large costs of adjustment affecting stocks of physical, human, and natural capital. Most economic analysis—and virtually all of the economic research that has been performed so far on the subject of climate change—is conducted in terms of comparative statics. It deals with economic equilibrium and the shift in equilibrium conditions that can be expected as a result of climate change. By contrast, the issues lying at the heart of climate change concern dynamics and disequilibrium—how long will it take for people to perceive changes in climate and respond to them? Will they refuse to acknowledge such changes when they occur, or will they quickly anticipate them? Will they adapt readily or with difficulty? Are there steps that can be taken to foster the recognition of change when it has occurred and accelerate adaptation to it—for example, by encouraging greater flexibility and reducing costs of adjustment? What is the scope for induced technical change that might lower the costs of both abating emissions and adjusting capital stocks? Answering such questions should receive a high priority in future research.

ACKNOWLEDGMENT

We are grateful to Peter Gleick and John Harte for conversations that have helped us better understand climate change and climate modeling.

NOTES

1. This property-value gradient is something that one could quantify using information from tax assessors' records, which is the method that Yohe employed.

2. The application of this methodology on a national scale to produce an overall estimate of the value of 4,000 square miles in the United States lost through flooding caused by a 0.5-meter rise in sea level is presented not in the EPA report but in Nordhaus's article. According to Ayres and Walter (1991), Nordhaus valued the flooded land at $2,023 per acre ($5,000 per hectare). However, there appears to be some confusion here: if one divides the $1.55 billion in table 1 by 4,000 square miles and assumes a capitalization factor of 10%, which Nordhaus appears to use for the other real estate calculations in table 1, the result is an imputed land value of approximately $6,000 per hectare. The figure of 4,000 square miles is from Nordhaus; the EPA report gives a range of 2,180 to 6,147 square miles for a sea-level rise of 50 centimeters (Ibid., p. 140).

3. Or other parts of the world. For example, Ayres and Walter (1991) point out that arable cropland is worth about $30,000 per hectare in the Netherlands. Since the United States is relatively abundant in land compared with many other countries, this makes it difficult to perform a simple extrapolation of damage estimates for global climate change from the United States to the rest of the world.

4. There can be a gap between marginal value and marginal cost because public infrastructure involves lumpy investment.

5. In a different but related vein, when Ayres and Walter (1991) estimate the costs of flooding in Third World areas threatened by sea-level rise, such as Bangladesh or the Nile Delta, they include in their estimate of resettlement costs an allowance for lost output over a 2-year period during which refugees from flooding are out of work (Ibid., p. 245).

6. There appears to be a discrepancy between the EPA data, reproduced in our table 2, and Nordhaus's data, reproduced in our table 1. The EPA estimates of <u>minus</u> $9.7 billion and <u>plus</u> $10.6 billion appear to become transposed into plus $9.7 billion and minus $10.6 billion. Note, also, that Nordhaus's point estimate of the impact on agriculture appears to be a loss of $0.45 billion.

7. See U.S. EPA (1989), pp. xlii and xliii. The discussion of health impacts in the Executive Summary is distinctly blander and more reassuring than the discussion in the main body of report.

8. They also suggest that the other costs of climate change are likely to be a good deal larger than the EPA/Nordhaus estimates.

WORKS CITED AND GENERAL REFERENCES

Ayres, R. U., and J. Walter. 1991. "The Greenhouse Effect: Damages, Costs and Abatement." *Environmental and Resource Economics,* vol. 1, no. 3, pp. 237-270.

Beckerman, W. 1991. "Global Warming: A Skeptical Assessment." Pp. 52-85 in D. Helm, ed., *Economic Policy Towards the Environment* (Oxford: Blackwell Publishers).

Booth, W. 1990. "Air Pollutant May Counter Global Heating." *Washington Post,* September.

Brown, L. R., and J. E. Young. 1990. "Feeding the World in the Nineties." Pp. 59-78 in L. R. Brown, A. B. Durning, H. F. French, C. Flavin, J. L. Jacobson, M. D. Lowe, S. Postel, M. Renner, L. Starke, and J. E. Young, eds., *State of the World 1990* (Washington, DC: Worldwatch Institute).

Cline, William R. 1991. "Scientific Basis for the Greenhouse Effect." *The Economic Journal,* vol. 101, pp. 904-919.

Daily, Gretchen C.; Paul R. Ehrlich, Harold A. Mooney, and Anne H. Ehrlich. 1991. "Greenhouse Economics: Learn Before You Leap." *Ecological Economics,* vol. 4, pp. 1-10.

Ehrlich, P. R., and A. H. Ehrlich. 1991. *Extinction* (New York: Ballantine).

Fisher, A. C., and W. M. Hanemann. 1986a. "Environmental Damages and Option Values." *Natural Resource Modeling,* vol. 1, pp. 111-124.

_____. 1986b. "Option Value and the Extinction of Species." *Advances in Applied Micro-Economics*, vol. 4, pp. 169-190.

_____. 1987. "Quasi-Option Value: Some Misconceptions Dispelled." *Journal of Environmental Economics and Management,* vol. 14, no. 2, pp. 183-190.

Gleick, Peter H. 1991. "The Regional Impacts of Global Climate Changes." Address, Peder Sather Symposium on Global Climate Change: European and American Policy Responses. Clark Kerr Campus, University of California at Berkeley, October 17.

Gleick, Peter H., and Maurer, Edwin P. 1990. *Assessing the Costs of Adapting to Sea Level Rise: A Case Study of San Francisco Bay.* Research Report (Stockholm: The Stockholm Environment Institute).

Haines, A. 1990. "The Implications for Health." Pp. 149-162. in A. Haines, ed., *Global Warming: The Greenpeace Report* (Oxford: Oxford University Press).

Harte, John. 1991. "What Is the Problem?" Talk, Peder Sather Symposium on Global Climate Change: European and American Policy Responses. Clark Kerr Campus, University of California at Berkeley, October 17.

Hoffman, J. S.; J. B. Wells, and J. G. Titus. 1986. "Future Global Warming and Sea Level Rise." Pp. 245-266 in G. Sigbjarnason, ed., *Iceland Coastal and River Symposium* (Reykjavik: National Energy Authority).

Intergovernmental Panel on Climate Change (IPPC). 1990. *Climate Change: The IPCC Scientific Assessment (Cambridge,* UK: Cambridge University Press).

Mendelsohn, Robert; William D. Nordhaus, and Daigee Shaw. 1992. "The Impact of Climate on Agriculture: A Ricardian Approach." Cowles Foundation Discussion Paper No. 1010 (New Haven, CT: Yale University, February).

National Research Council. 1978. *International Perspectives on the Study of Climate and Society* (Washington, DC: National Academy Press).

Nordhaus, William D. 1991. "Economic Approaches to Greenhouse Warming." Pp. 33-66 in R. Dornbusch and J. M. Potera, eds., *Global Warming: Economic Policy Responses* (Cambridge: The MIT Press).

Rosenberg, N. J.; W. E. Easterling, III, P. R. Crosson, and J. Darmstadter, eds. 1989. *Greenhouse Warming: Abatement and Adaptation* (Washington, DC: Resources for the Future).

Sundquist, E. T. 1990. "Long-Term Aspects of Future Atmospheric CO_2 and Sea-Level Changes." Pp. 193-207 in R. R. Revelle, et al., eds., *Sea Level Change* (Washington, DC: National Research Council, National Academy Press).

U.S. Department of Agriculture, Office of Management and Budget and National Climate Program Office. 1989. "Climate Impact Response Functions." Report (Coolfont, WV: Under the Auspices of IPCC).

U.S. Environmental Protection Agency (EPA). 1989. *The Potential Effects of Global Climate Change on the United States.* Report (Washington, DC: U.S. Congress).

Wilson, E. O. 1989. "Threats to Biodiversity." *Scientific American,* vol. 261, pp. 108-116 (September).

Weitzman, M. L. 1974. "Prices vs. Quantities." *The Review of Economic Studies,* vol. 41, pp. 477-491.

Yohe, G. W. 1989. "The Cost of Not Holding Back the Sea—Phase 1 Economic Vulnerability." In J. B. Smith and D. A. Tirpak, eds., *The Potential Effects of Global Climate Change,* Appendix B (*Sea Level Rise*) (Washington, DC: Office of Policy, Planning and Evaluation, U.S. EPA).

Contributors

James S. Clark
Department of Botany
Duke University
Durham, NC

Joel Darmstadter
Resources for the Future
Washington, DC

Anthony C. Fisher
Department of Agricultural and
     Resource Economics
University of California
Berkeley, CA

W. Michael Hanemann
Department of Agricultural and
     Resource Economics
University of California
Berkeley, CA

Stephen C. Peck
Electric Power Research Institute
Palo Alto, CA

Chantal D. Reid
Department of Crop Sciences and USDA
     Agricultural Research Service
North Carolina State University
Raleigh, NC

Norman J. Rosenberg
Battelle, Pacific Northwest Laboratories
Washington, DC

Thomas J. Teisberg
Teisberg Associates
Weston, MA

Michael A. Toman
Resources for the Future
Washington, DC

Paul E. Waggoner
Connecticut Agricultural
     Experiment Station
New Haven, CT

Gary W. Yohe
Department of Economics
Wesleyan University
Middletown, CT

155

Participants*

Workshop on Assessing Climate Change Risks:
Implications for Research and Decisionmaking

Resources for the Future, Washington, DC
March 23-24, 1992

Chris Bernabo
Science & Policy Associates, Inc.
Washington, DC

Michael Bowes
Office of Technology Assessment
Washington, DC

Hung-po Chao
Electric Power Research Institute
Palo Alto, CA

William Clark
John F. Kennedy School of Government
Harvard University
Cambridge, MA

William Cline
Institute for International Economics
Washington, DC

Chester Cooper
Battelle, Pacific Northwest Laboratories
Washington, DC

Rob Coppock
Committee on Science, Engineering,
    and Public Policy
National Academy of Sciences
Washington, DC

---

*Affiliations shown are those effective at the time of
the workshop.

Joel Darmstadter
Resources for the Future
Washington, DC

Roger Dower
World Resources Institute
Washington, DC

Hadi Dowlatabadi
Department of Engineering and
    Public Policy
Carnegie Mellon University
Pittsburgh, PA

Jae Edmonds
Battelle, Pacific Northwest Laboratories
Washington, DC

Anthony Fisher
Department of Agricultural and
    Resource Economics
University of California
Berkeley, CA

Robert Fri
Resources for the Future
Washington, DC

Michael Hanemann
Department of Agricultural and
    Resource Economics
University of California
Berkeley, CA

George Hidy
Electric Power Research Institute
Palo Alto, CA

John Jansen
R&D Department
Southern Company Services
Birmingham, AL

Sally Kane
National Oceanic & Atmospheric
    Administration
U.S. Department of Commerce
Washington, DC

Raymond Kopp
Resources for the Future
Washington, DC

Justin Lancaster
Environmental Science and Policy Institute
    and Harvard University
Cambridge, MA

Daniel Lashof
Natural Resources Defense Council
Washington, DC

Leonard Levin
Electric Power Research Institute
Palo Alto, CA

William Libro
Federal Governmental Affairs
Northern States Power Company
Washington, DC

Mack McFarland
E. I. du Pont de Nemours & Company
Wilmington, DE

Richard Morgenstern
Office of Policy Analysis
U.S. Environmental Protection Agency
Washington, DC

Peter Morrisette
Resources for the Future
Washington, DC

Daniel Newlon
National Science Foundation
Washington, DC

Stephen Peck
Electric Power Research Institute
Palo Alto, CA

Lou Pitelka
Electric Power Research Institute
Palo Alto, CA

Paul Portney
Resources for the Future
Washington, DC

Chantal Reid
Department of Botany
University of Georgia
Athens, GA

John Reilly
Economic Research Service
U.S. Department of Agriculture
Washington, DC

Lysbeth Rickerman
Office of Global Warming
U.S. Department of State
Washington, DC

Norman Rosenberg
Resources for the Future
Washington, DC

Michael Scott
Battelle, Pacific Northwest Laboratories
Richland, WA

Michael Springer
Office of Economic Policy
U.S. Department of the Treasury
Washington, DC

Raymond Squitieri
Council of Economic Advisers
Executive Office of the President
Washington, DC

Boyd Strain
Department of Botany
Duke University
Durham, NC

George Tolley
Department of Economics
University of Chicago
Chicago, IL

Michael Toman
Resources for the Future
Washington, DC

Paul Waggoner
Connecticut Agricultural
    Experiment Station
New Haven, CT

Gary Yohe
Department of Economics
Wesleyan University
Middletown, CT